AIRCRAFT ELECTRICAL SYSTEMS

SINGLE AND TWIN ENGINE

By J.E. Bygate

JEPPESEN.
Sanderson Training Products

Library of Congress Cataloging-in-publication number: 92-24372

JS312664B

Table of Contents

Preface

The purpose of this book is familiarization with the types of electrical systems that are in general use in light aircraft.

Included throughout are wiring schematics and excerpts from wiring schematics. These are from various makes and models of aircraft, and are for the purpose of illustration. In all cases the correct source of maintenance information is the airframe manufacturer.

To many people in the aviation maintenance field, the introduction of extra power system components presents a barrier as they have developed a large proportion of their expertise on the more numerous single engine aircraft types.

The aircraft electrical system is basic to any aircraft and differs little in concept regardless of size. As more engines are added, more components are added, and the basic systems are sometimes duplicated. The extra sophistication should not cause problems in understanding the concepts of the system.

Twin engine electrical power systems in the light general aviation aircraft types are becoming more numerous as the population of general aviation aircraft increases.

In the aviation industry, there is a tendency for aircraft to last a good while longer than the standard automobile so while it is necessary to have a good understanding of the more modern types of system, it is essential to maintain a good knowledge of the older type of system still operating in large quantities in the general aviation field. With this in mind, it has been found necessary to include in the text, descriptive material on twin generator systems as well as the more advanced twin alternator arrangements.

The text is prepared for the aircraft maintenance technician and it is assumed that the user has a basic understanding of electrical fundamentals such as electrical units, ohms law, use of measuring instruments, magnetism, electromagnetism, and electromagnetic induction.

AIRCRAFT ELECTRICAL SYSTEMS

PART I
SINGLE ENGINE

INTRODUCTION

In all aircraft electrical installations, the whole system is really divided up into several subsystems which are: primary or battery power, generated power, power distribution or power consuming circuits, and auxilliary power. Both the primary and generated power are applied to the main electrical busbar.

To be sensible, safe and viable, an aircraft electrical system must have the following four characteristics:

1. A primary power source (the battery) which is controllable by the pilot(s) and which will allow sufficient duration for the operation of electrical equipment should the engine driven generator fail.

2. An engine driven generation device which is automatically controlled as to output with sufficient output to supply all normal loads and charge the battery and is controllable by the pilot(s).

3. An electrical distribution system which distributes power to the various circuits through appropriately rated protection and controlling devices.

4. Optional auxilliary power capability which should be arranged so as to be available even though the aircraft battery may be dead.

NOTE: Some operations require that their aircraft are equipped with an alternative power capability from an external source, especially where severly cold climatic conditions or high frequency short duration flights are normal or prevalent.

A. Electrical Distribution System

The distribution system starts at the busbar and contains all the necessary circuitry and components to operate lighting. Flaps, landing gear indicators, motors, radios, and secondary power such as 115V, 400 Hz AC for specific tasks.

1. Distribution system components

The electrical distribution system is comprised of the following common types of components:

a. Electrical busbars

In practice these busbars are strips of copper with holes drilled in them at appropriate intervals which are connected to one side of a row of circuit breakers and to the incoming battery and generator power line.

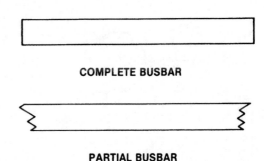

COMPLETE BUSBAR

PARTIAL BUSBAR

b. Circuit breakers

Circuit breakers are devices which will remove the power from a component if an overload is present in the component and its circuitry. A circuit breaker is connected in series with the circuit it is to protect it. When the overload is removed, the circuit breaker can be reset. Automatic reset type circuit breakers are not used in aircraft.

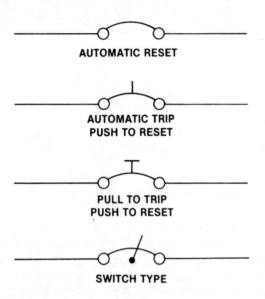

AUTOMATIC RESET

AUTOMATIC TRIP
PUSH TO RESET

PULL TO TRIP
PUSH TO RESET

SWITCH TYPE

c. Transistors

Transistors are semiconductor diodes with three electrodes. They transmit an electrical signal across a resistor.

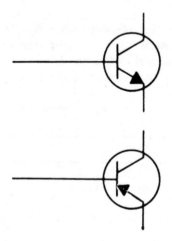

d. Resistors
They have electrical resistance and are used in an electric circuit for protection, operation, or current control.

e. Diodes

They are electron check valves that allow the flow of electrons in one direction but not in the opposite direction.

f. Fuses

Fuses perform basically the same function as a circuit breaker except that the fuse has a fine wire link which fractures on overload so the fuse must be replaced when the overload is traced and removed.

g. Switches

Switches are used to control the availability of power to the components. There are various arrangements of switch contacts depending on the operation required by the circuit.

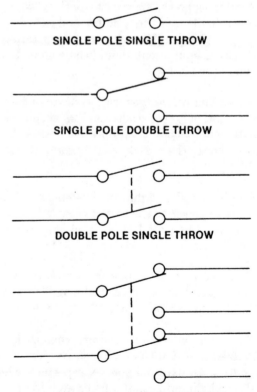

SINGLE POLE SINGLE THROW

SINGLE POLE DOUBLE THROW

DOUBLE POLE SINGLE THROW

DOUBLE POLE DOUBLE THROW

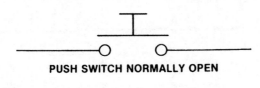

PUSH SWITCH NORMALLY OPEN

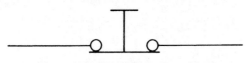

PUSH SWITCH NORMALLY CLOSED

h. Lamps

Various types of lamps are used both internally and externally for various functions. Those used for illumination are manually operated, while indicator lamps are controlled by a component.

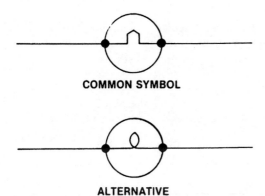

COMMON SYMBOL

ALTERNATIVE

i. Relays

Relays are remote controlled electromagnetic switches. They not only provide control, but also reduce long lengths of heavy wire.

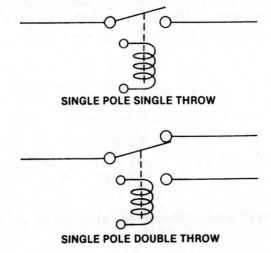

SINGLE POLE SINGLE THROW

SINGLE POLE DOUBLE THROW

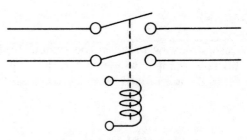

DOUBLE POLE SINGLE THROW

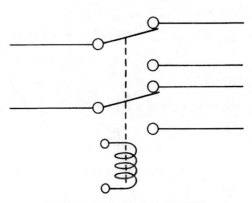

DOUBLE POLE DOUBLE THROW

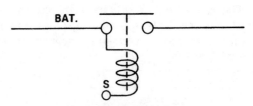

**HEAVY DUTY CONTACTOR
(MASTER RELAY)**

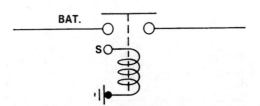

**HEAVY DUTY CONTACTOR
(STARTER RELAY)**

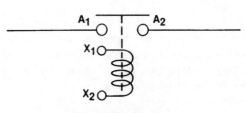

**HEAVY DUTY CONTACTOR
(AN TYPE)**

5

j. Motors

Motors may be found in the system to drive fuel booster pumps, windshield de-icing pumps and wing di-icing pumps as some examples.

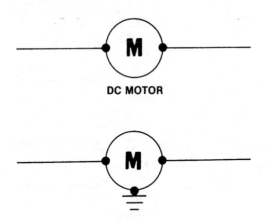

DC MOTOR

DC REVERSIBLE MOTOR

k. Actuators

The use of actuators for trim control, flap drives and cowflaps is becoming more prevalent as is actuator control of landing gear systems.

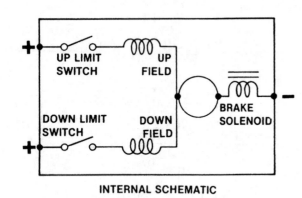

INTERNAL SCHEMATIC

l. Plugs

Male fittings for making electrical connections by insertion in a receptacle or body of electrical equipment to a circuit. A device for connecting electric wires to a jack.

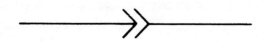

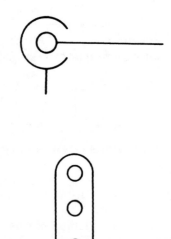

m. Heating elements

Electrical heating elements are used extensively in pitot tubes and propeller de-icing as well as in windshield heat systems.

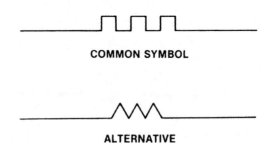

COMMON SYMBOL

ALTERNATIVE

n. Solenoids

Solenoid operated valves are used in hydraulic control and selection while pure solenoids may be used to lock landing gear control handles.

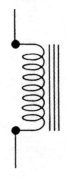

o. Radios

A means of communication with the ground. They are installed in various numbers and types. Fully instrumented aircraft may have at least four radios of various types.

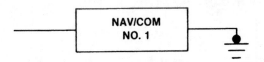

LETTERING IN BOX DETERMINES TYPE OF RADIO

p. Inverters

As radio navigational equipment becomes more prevalent, the need for inverters to power syncro transmitters and indicators rises.

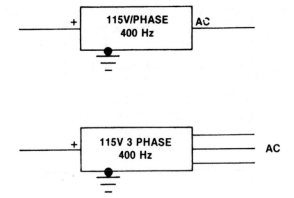

q. Microswitches

Microswitches are used in landing gears, flap controls, and indication circuits. Microswitches have similar arrangements to normal switches, i.e., S.P.S.T., S.P.D.T., D.P.D.T., depending on their application.

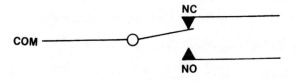

2. Schematic diagram

Reading a schematic diagram of a power distribution circuit is not much different from reading a good road map. The major difference is in the symbols an location identification.

Most schematics use a common set of symbols to represent various devices. These symbols are seldom labeled with the name of the device. A good working knowledge of these symbols, and the ability to recognize them on sight is, therefore, vital to understanding circuit diagrams.

By studying the preceding drawings you will notice that most of the devices can be easily recognized and their function clearly understood.

Note: See the schematic diagram example on next page.

Review Questions:

1. What purpose does the battery serve in the aircraft electrical system?

2. An engine driven generator is used to supplement the battery. Why?

3. Auxilliary power is used under what typical conditions?

4. What is the purpose of electrical busbars?

5. What type of circuit breaker is not used in aircraft electrical circuits?

6. Describe the inconvenience of fuses in a circuit.

7. What is the electrical differences between a starter relay and a master relay?

8. What is the purpose of an actuator?

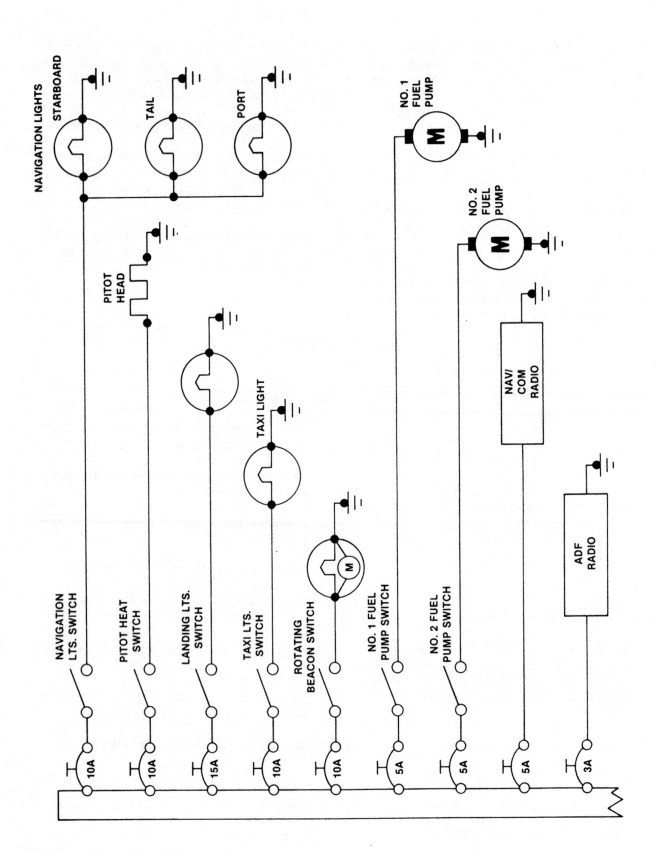

CHAPTER 1
Single Engine Electrical System

A. Battery Control Circuit Block

1. Operation

Selection of the battery control to "on" connects the battery to the busbar through the battery drain monitor. Any loads connected to the busbar will be supplied with current from the battery. Selection of the battery control to "off" disconnects the battery from the busbar and isolates all loads from the battery.

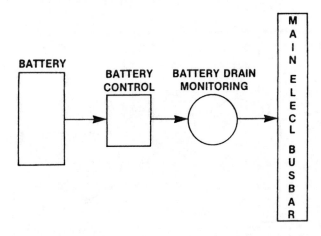

Fig. 1-1 Battery control circuit block schematic

B. Battery Control Circuit Wiring

In actual fact, the battery control circuit consist of: (1) the battery, (2) the battery relay—an electromagnetic switch whose main contacts when closed, connect the battery output to the busbar, (3) a battery master switch to control the battery relay "on""off" function, (4) an ammeter which shows how much current is drawn from the battery by the loads, and (5) the electrical busbar, a conveniet connection point to which the loads may be connected.

1. Operation

The battery voltage is available at terminal "B" on the battery relay. When the battery master switch is closed, current flows through the relay coil, the "S" terminal and the master switch to ground. The current through the coil causes the relay contacts to close through electromagnetic attraction. This allows battery output to go through the relay contact B_1 to the busbar through the ammeter.

Power is now available to the electrical busbar and will supply current to any circuits on the busbar which are selected "on." It will be noted that power is also available to the wire marked "to start relay" for use when starting the engine.

Selecting the battery master switch to "off" de-energizes the battery relay. Its contacts spring open and battery power is disconnected from the busbar and starter relay.

C. Battery Control and Starter Control Circuit

This circuit has added the following components to the battery control circuit seen in Fig. 1-2. The added components are: (1) a starter relay, (2) a starter motor, (3) a spring loaded to "off" starter switch, and (4) an 8 amp circuit breaker.

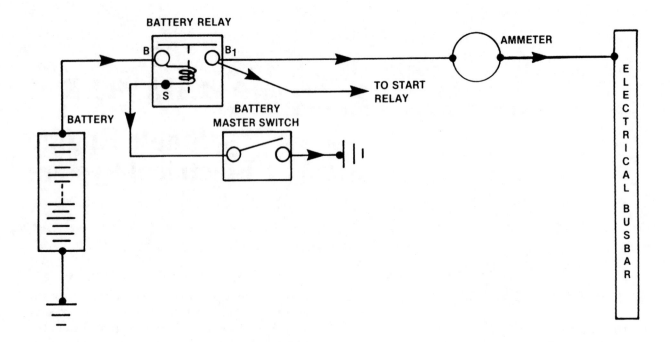

Fig. 1-2 Battery control circuit wiring schematic

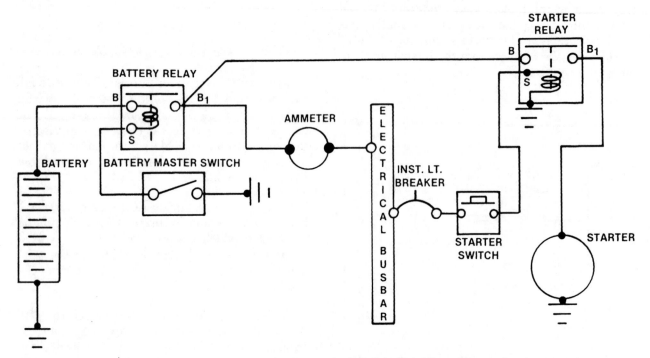

Fig. 1-3 Battery and starter control circuit wiring schematic

1. Operation

Selection of the battery master switch to "on" energizes the battery relay and supplies power to the busbar and to the "B" terminal of the starter relay. As power is available at the busbar, power is also available through the INST LT breaker to the starter switch.

10

Depressing the starter switch allows power through to the "S" terminal of the starter relay. Current flows through the starter relay coil to ground and energizes the starter relay. The starter relay contacts close and power is fed across the contacts B and B_1 to the starter motor turning the motor. When the engine fires and starts, the starter switch is released. This de-energizes and opens the starter relay contacts and the starter motor stops.

D. Generator Control Circuit

The generator provides electrical power when the engine is running to charge the battery, and supply the loads which are connected to the busbar. The generator and control circuit operate as follows:

When the engine is started, the generator starts producing electrical output. As soon as the output of the generator exceeds that of the battery, the generator control switches the generator output to the busbar. If the voltage of the generator rises above or falls below the set regulated

level the control system automatically adjusts the voltage back to normal. If current drawn from the generator exceeds rated output, the control unit automatically limits output current to max rated output. When the engine is shut down the control unit automatically disconnects the generator from the busbar to prevent the battery from trying to motor the generator.

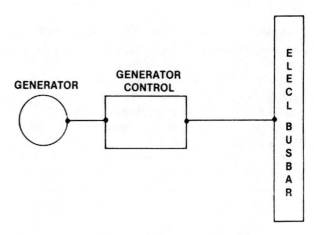

Fig. 1-4 Generator control circuit block schematic

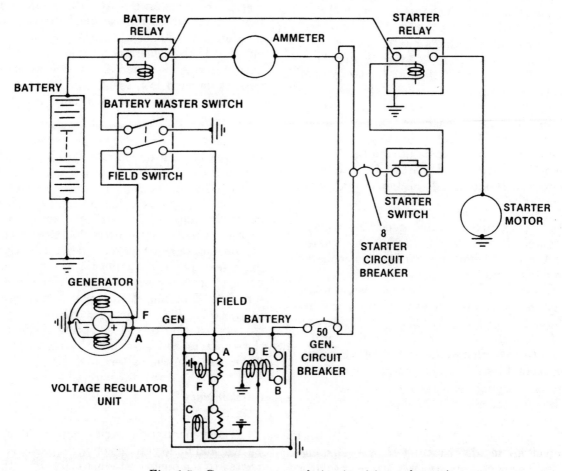

Fig. 1-5 Generator control circuit wiring schematic

11

Seen in Fig. 1-5 is the basic battery control circuit with the generator and its control added. The generator circuit consists of: (1) a shunt wound DC generator, (2) a standard three unit regulator, (3) a 50 amp circuit breaker, and (4) a field switch which is mechanically interlocked to the battery master switch.

1. Operation

Selection of the battery master switch to "on" energizes the battery relay which closes its contacts and connects battery power through the ammeter to the busbar and from the output of the battery relay to the input of the starter relay. (The field switch for the generator is closed when the battery master switch was selected "on.") With the ignition switches "on," depressing the starter switch will allow power to be applied to the starter relay coil. This energizes the starter relay closing its contacts and connects power to the starter motor which turns the engine allowing it to start.

As the engine starts, the generator armature turns inside the poles of the field system. Residual magnetism in the field produces a small electrical output which flows from the generator and through the field windings out of the "F" terminal through the field switch into the regulator unit "field" terminal and through the two sets of normally closed contacts marked A to ground in the regulator base. This current further excites the generator and generator voltage output builds up rapidly. The generator output voltage is applied to the regulator unit through the "GEN" terminal through the coil marked C to the contacts marked B and can go no further until contacts B close.

At approximately 12.6V generator voltage the contacts B close due to 12.6V being applied across the voltage sensing coil D. This connects the generator power through the series coil E through the contacts B and the generator circuit breaker to the busbar allowing power from the generator to supply busbar loads and drive current through the ammeter, through the battery relay contacts to charge the battery. (Coils D and E are wound on the same former and under generator charging conditions, coil E assists coil D in keeping the contacts B closed.)

The generator voltage continues to rise and at the voltage regulator setting of 14.0 volts, coil F comes into operation. As the voltage goes over 14 volts, the coil F energizes and opens the normally closed contacts A. This forces the field current to go through the resistor connected across contacts A and will decrease the field current. Less field current means less field excitation and less generator voltage output.

As generator voltage decreases, this is sensed by coil F which at less than 14 volts, can not hold the contacts open. The contacts close, shorting out the resistor. This allows increased field current through the closed contacts which causes increased excitation and the generator output rises back to 14 volts. This process repeats itself many times each second and maintains the voltage output at a mean level of about 14.0 volts.

With the voltage being regulated at 14 volts, the generator must be projected against accidental or intentional overloading. This is achieved by the use of the coil C and its relevant contacts and resistance "L." The coil C carries full generator output current and is calibrated so all current flowing from the generator to the loads up to 50 amps, has no effect on the contacts "L." If current exceeds 50 amps, contact "L" will open. This inserts the resistance across the contacts in the field circuit which reduces field current and excitation which in turn, reduces generator output voltage. If the generator voltage is reduced, its output current is reduced. As long as excessive current demand is present, the contacts remain open and voltage is depressed. This limits the current to a safe level.

The generator must be protected from battery voltage during engine closedown. This is achieved by the action of coil E and contacts B during this operation. As the generator runs down, its output voltage decreases. When the generator voltage becomes less than the battery voltage, current flows from the battery through the busbar, circuit breaker and coil E back to the generator. The current flowing through coil E is now going in a reverse direction to previous current flow. This causes an opposite magnetic field to that of coil D. When coil E magnetic effect cancels out that of coil D, spring tension opens the contacts and automatically disconnects the generator from the battery.

2. Summary

The generator is excited by residual magnetism and builds up its voltage rapidly. It is connected to the busbar automatically when the generator voltage exceeds the battery voltage by the closing action of the *reverse current cutout.*

The voltage output of the generator is controlled or regulated by the *voltage regulator.* The current output is limited to rated output by the *current limiter.* The generator is automatically disconnected from the battery on engine close down to prevent damage to the generator by the reverse current function of the *reverse current cutout.*

Review Questions:

1. What purpose does the battery relay serve?

2. How is the ammeter connected in the circuit between the battery and busbar?

3. What is meant by de-energize when applied to a relay?

4. The starter switch has a special function. Name it.

5. What is meant by energizing the starter relay?

6. The starter control circuit is protected by a circuit breaker. What other circuit is protected by the same circuit breaker?

7. What is the purpose of the generator?

8. How is the generator field switch connected in the circuit?

9. How does the generator receive its initial excitation?

10. What component switches the generator onto the busbar?

11. Two devices protect the system from electrical overload. What are these devices?

12. Explain the principle of operation of the voltage regulator.

13. How is the generator protected from the battery when the engine is shut down?

14. Explain how the current limited circuit operates.

CHAPTER 2

Single Engine Power Distribution System

The aircraft electrical distribution system contains the following basic components:

1. *The electrical busbar* which provides a convenient connection point for all aircraft electrical system loads with the exception of the main starter system loads with the exception of the main starter current. The busbar is the main interface point between the two basic power sources (battery and generator) and the aircraft electrical loads.

2. Connected to the busbar are *circuit breakers* or *fuses* to protect each of the circuits which are connected as electrical loads to the system.

3. Connected between each circuit breaker and its equipment are *control switches* which, together with the protection devices, are rated so as to safely take the individual current required for that load.

4. *Wiring* between the busbar, protective devices, switches and their loads, is rated in each case to safely carry the particular amount of current required for each particular load.

5. The *loads* are the electrical consuming devices connected to the controlling switches are, of course, such equipment as lights, motors, relays, electrical instruments, heaters, and avionics units such as nav/com radios, ADF, transponder, DME, and weather radar. Each load, when operating, is connected to the busbar in parallel from the busbar to ground.

A. Operation

With all switches off, the first event that happens is selection of battery power placing the battery master switch "on." Power from the battery is applied to the busbar and to the input terminal of the starter relay.

Selection of the starter control switch "on" applies power to the starter motor and the engine starts. When the engine starts, the generator excites and the generator power is automatically applied to the busbar. The voltage of the generator is higher than battery voltage and this difference causes some generator output current to go through the battery for charging.

The loads that can be switched onto the busbar are calculated so as to not exceed the output capability of the generator; therefore, all loads switched on during the condition of the generator supplying the busbar are taken care of by the generator output. Selection of the appropriate circuit control switch will apply power to the particular load selected and cause that particular device to function. As more circuit control switches are selected, more of the load components will function. This increases the current drain from the generator.

If the generator is switched "off" or "fails" the distribution system is supplied with power from the back-up source which is, of course, the battery. If this happens, it may be necessary to switch off non-essential loads to conserve battery power as it is no longer being charged by the generator.

15

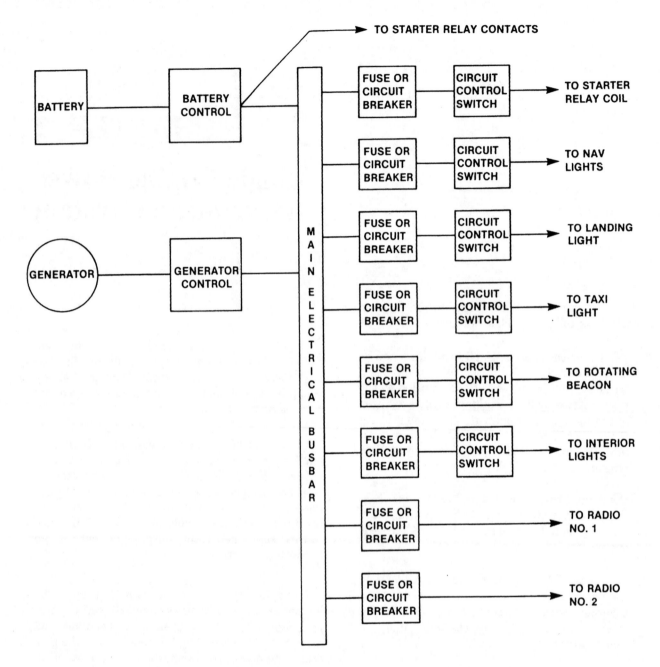

Fig. 1-6 Power distribution system block schematic

B. Split Busbars

In later aircraft it has been a common design practice to isolate transistorized equiment from the main electrical busbar during periods of violent variations of voltage on main busbar during starting, or when using external power for starting.

Initially, this was achieved by connecting the electronic equipment protective devices to a separate electronic busbar and connecting the electronics busbar to the electrical busbar through an isolating switch which was usually named the avionics master switch. The avionics master switch should, of course, be "off" during starting. This isolates the avionics equipment from the main busbar.

A later arrangement of splitting busbars or isolating susceptable avionics equipment from transient voltages during starting or when

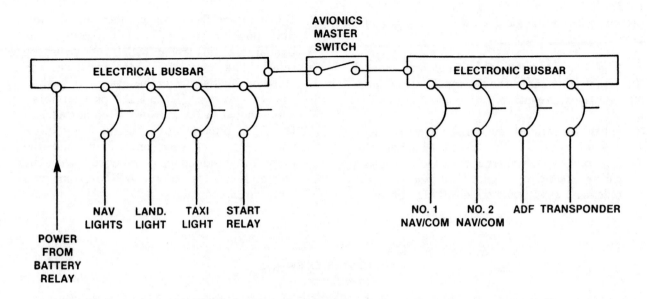

Fig. 1-7   Earlier split busbar arrangement with an avionics master switch.

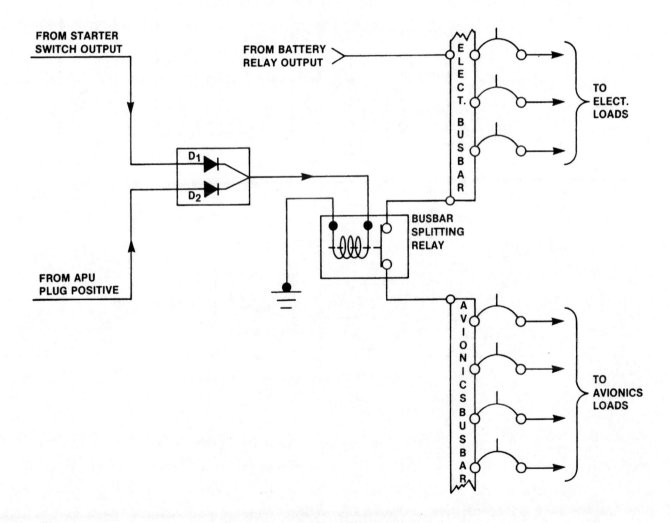

Fig. 1-8   Busbar splitting relay schematic

ground power is plugged in, was by the use of a busbar splitting relay.

The relays normally closed contacts connect the avionics busbar to the electrical busbar under all conditions, except:

1. When the starter switch is selected "on", power is applied to the coil of the busbar splitting relay which energizes the relay causing the contacts to open. This isolates the avionics busbar from the electrical busbar.

2. When power from the auxillary power unit plug positive is applied to the coil of the busbar splitting relay which energizes the relay. The contacts open isolating the avionics busbar from the electrical busbar. The two diodes, D_1 and D_2, prevent interaction between the starter switch output and the APU plug positive circuit. They block current flow between one circuit and the other, but conduct to the relay coil when the starter is selected or the APU is connected and switched on.

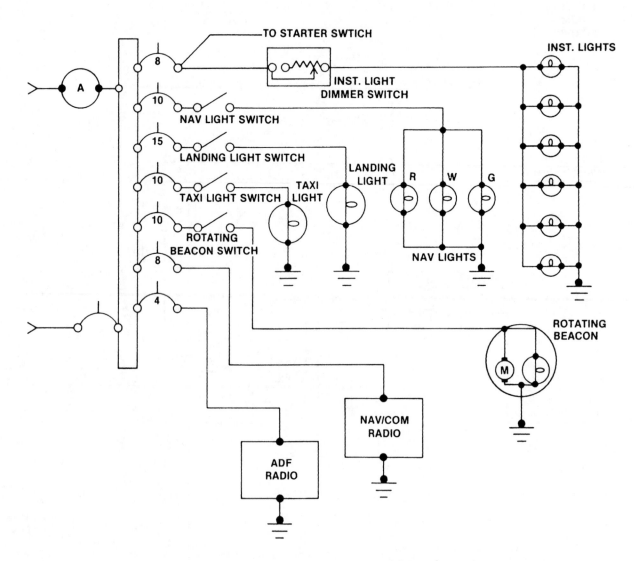

Fig. 1-9 Power distribution system wiring schematic

18

Shown in Fig. 1-9 is a typical light aircraft distribution system. Each load is connected to the busbar through its circuit control switch and circuit breaker. The circuit control switches for the two radios are mounted on the radio front panel and are not shown in Fig. 1-9.

Occasionally circuit breakers may supply more than one load as shown on the instrument light breaker which also supplies the starter switch and starter relay coil. Some aircraft systems have more loads connected. When this happens, additional circuits breakers and circuit control switches are connected to the busbar. Typical extra loads which may be connected are an electrical turn and slip indicator, electric gyro horizon, a flap motor, a landing gear motor and indication, and a rotary or static inverter to supply alternating current for instruments such as the gyrosyn compass. The type of loads will, of course, vary with aircraft make and model, and the particular role of the aircraft.

Review Questions:

1. What is meant by electrical busbar?

2. How is the wiring rated in each of the load circuits?

3. What is the purpose of a circuit breaker?

4. What is meant by split busbars?

5. How does a split busbar relay protect avionics equipment?

6. Where would you find the control switches on radio equipment?

CHAPTER 3
Single Engine Secondary Power Supplies

A. DC Current to AC Current

On some aircraft systems there is a requirement for alternating current (AC) to supply power to operate alternating current equipment. In light aircraft where the primary source of power is direct current (DC), it is necessary to convert the DC power to AC. The most common way of accomplishing this is to use an inverter. An inverter is a device which converts direct current (DC) to alternating current (AC).

Aircraft AC electrical supplies have a frequency output of 400 Hz. This particular frequency was chosen as a standard as it allows the design of lighter alternating current components than would be possible with lower frequencies.

B. Rotary Inverter

A rotary inverter consists of a DC motor which is powered from the DC busbar mechanically coupled to an AC generator. When the motor is switched "on," it drives the AC generator which produces alternating current and feeds the alternating current to the AC busbars. The instrumentation which requires AC current is connected through suitable fuses and switches to the AC busbars.

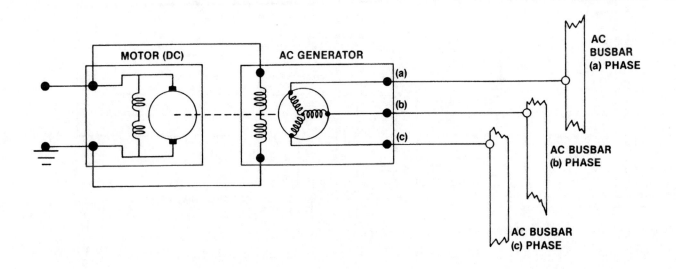

Fig. 1-10   Rotary inverter schematic

C. Inverter Outputs

The alternating current supplied by an inverter has the following output characteristics: 115V AC, 400 Hz or 26V AC, 400 Hz. The output from the inverter may be single phase or three phase, depending on the loading requirements of the system. In most light aircraft the requirement is 26V AC, 400 Hz.

The schematic in Fig. 1-10 shows the concept of a rotary inverter. Each inverter comes equipped with circuitry which controls the motor speed and so controls the frequency to 40 Hz and also with a means of controlling the generator field excitation to control the output voltage to 115V or 26V AC.

D. How Inverters Operate

1. Typical 3-phase rotary inverter

Power is applied to the aircraft DC busbar by placing the battery master switch to "on." The engine is started, bringing the aircraft generator on line.

Selection of the inverter on/off switch to "on" starts the inverter motor running. When the motor runs up to speed, it is controlled at that speed by the adjustable resistor **A** which is set so that sufficient motor field excitation is allowed through the field **B** to allow the motor speed to produce 400 Hz ± 20 Hz at the regulated voltage of the aircraft. In this type of system, the frequency will be low when the aircraft generator is not feeding the busbar.

As the motor runs at a relatively constant speed, it drives the AC generator which produces an output of around 115V AC 3-phase to the AC busbars (a), (b), and (c). This voltage is controlled to the 115V level by presetting the adjustable resistor **D** to give the correct excitation current to the AC generator field **C** at the regulated voltage level set for the aircraft. The AC voltage will be low if the aircraft generator is not connected to the DC busbr.

The low frequency and voltage situation is usually eliminated by ensuring that the inverter is not selected until after the engine is started, as the inverter motor load can be quite heavy and

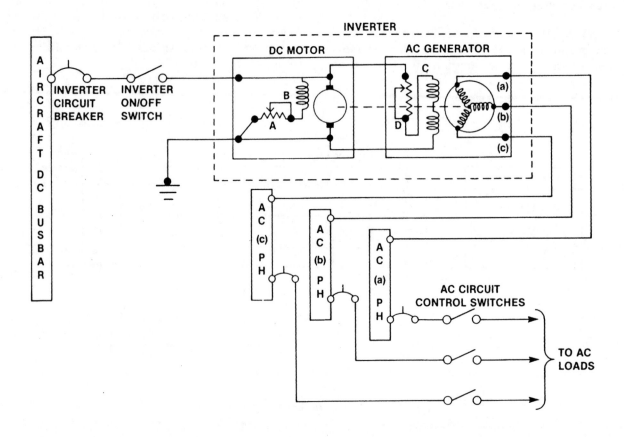

Fig. 1-11 Schematic of a typical 3-phase aircraft rotary inverter with controls.

22

would drain the battery during the starting operation. Once the inverter is running, the AC loads can be selected by using the appropriate AC circuit control switches.

2. Typical 26V, 400 Hz single phase inverter

The system in Fig. 1-12 is controlled and operates the same as described in Fig. 1-11. The only difference is it is a single phase output of 26V AC, 400 Hz. The single phase output is connected between the AC busbar and aircraft electrical ground.

3. Solid-state static inverter

A solid-state inverter consists of a solid-state sinusoidal oscillator and a 400 Hz transformer. It operates by placing the inverter control switch to "on," allowing 12V DC from the busbar to supply the sinusoidal oscillator. The sinusoidal oscillator immediately starts up, and by transistor switching action, the oscillator produces 12V sine wave AC, 400 Hz at the primary of the transformer.

The transformer steps up the 12V AC, 400 Hz in the primary winding to 26V AC, 400 Hz in the secondary winding. This 26V AC, 400 Hz is fed to the AC busbar as inverter output.

NOTE: The frequency and voltage controls are built in to the sinusoidal oscillator and usually have no external adjustment. Troubleshooting generally consists of checking DC input and AC output with a multimeter for correct value.

Review Questions:

1. What is meant by secondary power supplies?

2. What is the purpose of an inverter?

3. Why do we use 400 Hz as a frequency standard on AC systems?

4. How is the frequency controlled in a rotary inverter?

5. What method is used to control voltage in a rotary inverter?

6. Under what conditions would a rotary inverter, as described in the text, be operating at the correct frequency and voltage?

7. Solid-state inverters contain two basic elements. What are they?

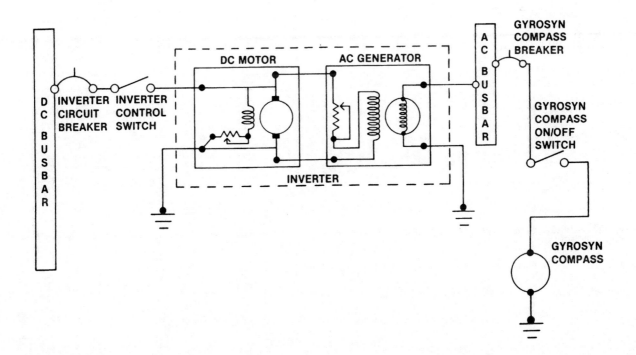

Fig. 1-12 Schematic of a typical 26V, 400 Hz single phase inverter

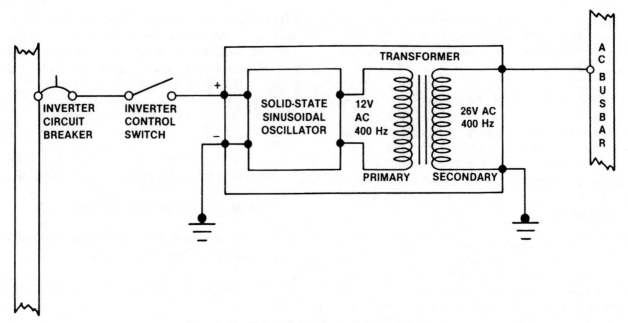

Fig. 1-13   Schematic of a solid-state inverter

CHAPTER 4

Single Engine Wiring Considerations For Power and Distribution Circuits

A. Wiring Considerations

It can safely be assumed that all light metal skinned aircraft are designed to use the metal structure of the airframe as the ground returns path for all major electrical circuits. This procedure of using the airframe as ground return is to reduce the amount of wire used in the electrical system which in turn saves weight.

Each circuit function requires a specific current in order to function correctly so it is necessary to wire each circuit so that the wire will adequately handle the current that flows in that circuit with a minimum of volts drop or loss over the circuit's wire length.

The positive output of the power sources is connected to the main electrical busbar and the ground return is connected to the negative side of the power sources. This ensures that any load connected between the busbar and ground will be connected in parallel across the power source.

B. Wire Sizes

1. Battery Control Circuit and Starter

The wires marked **A** will all, at some time or other, have to carry starter motor current which could be in the order of 175 to 250 amps, depending on the engine size. The wire gauge for these wires would be in the range of 2 AWG to 00 AWG.

The wires marked **B** are the main power feed to the busbar from the battery and have to supply all loads connected to the busbar. These loads generally would total between 30 and 60 amps. The wire gauge for these wires coulid range from 10 AWG to 6 AWG.

The wires marked **C** carry the current required to energize the coils for the master and starter relays. These currents are in the 2 to 5 amp range and would be wired with 20 AWG to 18 AWG wire.

2. Generator system

The wires marked **A** are the main generator output to the busbar and the current could be 30 to 60 amps. These wires would be in the 10 to 6 AWG range.

The wires marked **B** are carrying field current of 2 to 2½ amps. 20 AWG would be sufficient for these wires.

3. Distribution system

In the distribution system, the wire sizes are naturally determined by the amount of current flow through each particular load. As each load is individually protected by circuit breakers or fuses, this is an added safety factor.

Particular attention should be paid to the maintenance manual wiring diagrams when selecting wire for replacement in any circuit in the aircraft system. If the correct gauge is unavailable, the safest procedure is to use the next larger gauge, i.e., for 18 use 16 AWG.

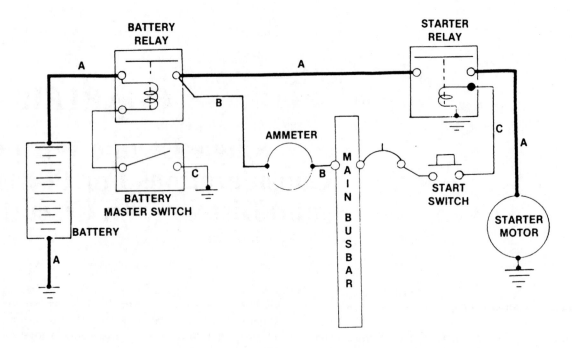

Fig. 1-14 Schematic of the battery control circuit and starter showing wire sizes.

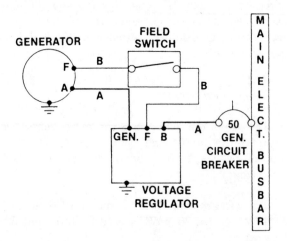

Fig. 1-15 Schematic of the generator system and control wire sizes.

Review Questions:

1. Name two reasons that most aircraft use the airframe as ground return for the electrical system.

2. Explain why it is necessary to use No. 10 wire for generator output and No. 20 wire for the field circuit.

3. What gauge wire would you use if the wire you want to replace is out of stock?

CHAPTER 5

Single Engine
Auxiliary Power Systems

A. Auxiliary Power Supply Systems

Some operators, because of the type of operation, require provisions on their aircraft for an auxiliary or external power source to be plugged into the aircraft for starting or ground servicing. This is especially true for severely cold climatic conditions or operations which have repetitive short duration flights. By using external power, the aircraft's battery is conserved.

It is arranged that a suitable external power plug is located conveniently on the aircraft and the power is connected to a system point which is usually connected to the battery master relay output terminal.

B. System Operation

External power of the correct voltage and polarity is applied to the external power plug. The positive terminal provides power up the wire marked **A** to the output terminal of the battery relay.

Power to operate the starter and any of the load circuits is available through the aircraft wiring from this point even if the battery relay is not closed. This provides an alternate source of power to start the aircraft should the aircraft battery be flat.

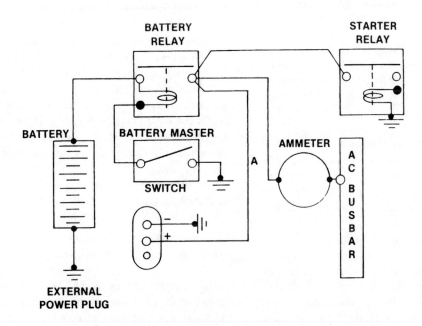

Fig. 1-16 Simple auxiliary power system schematic

27

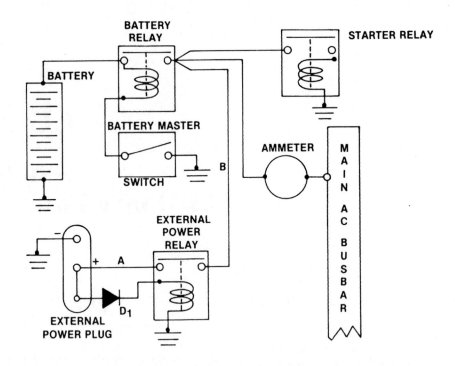

Fig. 1-17 Schematic of a relay controlled auxiliary power system

C. Relay Controlled Auxiliary Power System

The system in Fig. 1-17 differs from the system in Fig. 1-16 in that it has a polarized relay which connects the external power positive line to battery relay output.

With the battery master switch "off," the auxiliary power unit (APU) is connected to the external power plug. When the auxiliary power is switched "on," current flows through the positive pin to the link to the small pin through diode, D_1, to the coil of the external power relay and to ground. This energizes the external power relay and its contacts close.

Power is now allowed through the external power plug positive through the external power relay contacts on wires **A** and **B** to the battery relay output contact and from there to the starter relay input and the main busbar. Selection of any load on the main busbar will cause the load to be supplied by the auxiliary power unit.

If the incorrect polarity is connected to the positive pin of the external power plug diode, D_1, will block the negative polarity and the external power relay will not be energized. This protects the system against the incorrect polarity being connected to the aircraft electrical system.

Caution: On many light aircraft the external power plug is the standard N.A.T.O. 3-pin type regardless of system voltage. Care should be taken to ensure that the correct voltage APU is used on particular aircraft. Double check on aircraft system voltage before connecting the APU.

1. Battery charging system from an APU

More modern single engine systems are being equipped with circuitry which enables the aircraft battery to be charged when an APU is plugged into the aircraft. This is achieved by the use of a battery relay closing circuit connected across the contacts of the battery relay. If the aircraft battery voltage is too low to close the battery relay, an alternative supply is available through the battery relay closing circuit from the APU to close the battery relay and allow the APU power through to charge the battery.

With the APU connected correctly and switched on power is available through external power relay and wires **A** and **B** to the battery relay output terminal. If the battery has insufficient voltage to energize the battery relay, the battery relay closing circuit board routes a reduced voltage through its diode, resistor, and fuse to the battery relay input terminal.

28

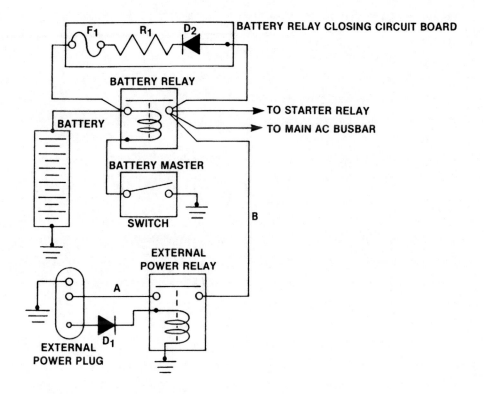

Fig. 1-18 A battery relay closing circuit schematic

If the battery master switch is selected "on," current flows through the battery relay coil to ground which energizes the battery relay from the closing circuit voltage. The battery relay contacts close and full APU power is allowed through the contacts to charge the battery.

The diode, D_2, prevents the battery from bypassing the battery relay when the battery relay is open with no APU connected. The fuse, F_1, will blow if the battery is shorted when APU power is applied to the system. The resistor, R_1, drops the voltage applied to the battery relay input terminal.

Caution: When charging the aircraft battery, it is essential that the APU is well regulated and controllable with regard to voltage and current capability.

Review Questions:

1. Explain some of the problems with early APU systems in aircraft.

2. How is the relay controlled auxilliary power circuit protected from reverse polarity?

3. What precautions should be taken before connecting an APU to the aircraft?

4. What is the purpose of the diode, D_2, and the fuse, F_1, in the battery relay closing circuit?

CHAPTER 6
Single Engine Automotive Sytle Alternator Systems

The DC generator installed on light aircraft up until around 1964 had the following shortcomings: (1) its power to weight ratio was poor, (2) its performance at low RPM was minimal, and servicing and maintenance was high because of high current carrying characteristics of brushes and commutator.

Around 1964 light aircraft manufacturers began installing automotive style alternators in their aircraft to alleviate the shortcomings of the DC generator system. The type of alternator installed was a rotating field device which has its output rectified to DC by a diode rectifier mounted on its case. It is not, therefore, a true alternator such as we may see on larger commercial aircraft.

A. System Description

The alternator system contains the following components: (1) the alternator and (2) the voltage regulator.

1. Alternator

The alternator is comprised of a rotor, a stator and a rectifier. The *rotor* is the rotating field coil supplied through brushes and slip rings with excitation current from the aircraft busbar. The *stator* is the output winding inside which the rotor rotates. The stator coils are wound 120° apart and its output is 3-phase AC which is fed to the *rectifier*. The rectifier is six diodes whose purpose is to change the 3-phase AC output from the stator into DC output external to the alternator.

2. The voltage regulator

The voltage regulator is comprised of a *voltage sensing coil* **A** to sense busbar voltage, a *moving contact* **B** which is moved by the voltage sensing coil, a *resistor* **C** which is inserted or removed by the voltage sensing coil operating the contacts, and *fixed contacts* **D** and **E** whose purpose is explained in the text.

B. Alternator Operation

The battery relay system is the same system as discussed in Chapter 2. Power is applied to the main busbar by selecting the battery master switch "on." The alternator field switch which is interlocked with the battery master switch is also placed "on."

Power now feeds the busbar through the battery relay contacts and continues from the busbar through the alternator field switch, the normally closed contacts **D** and **B** the "F" terminal on the regulator to the "F" terminal on the alternator then through the rotor winding to ground. The field is now energized by power from the busbar and a field current of approximately 2 amps is flowing through the field rotor coils. The engine is started.

As the rotor turns inside the stator coils, the rotor magnetic field induces an alternating voltage in the stator windings. The voltage builds up rapidly to approximately 14V. The rectifier changes the AC output from the stator into DC output at the B+ terminal of the alternator. This DC voltage is fed to the main bus through the 60

31

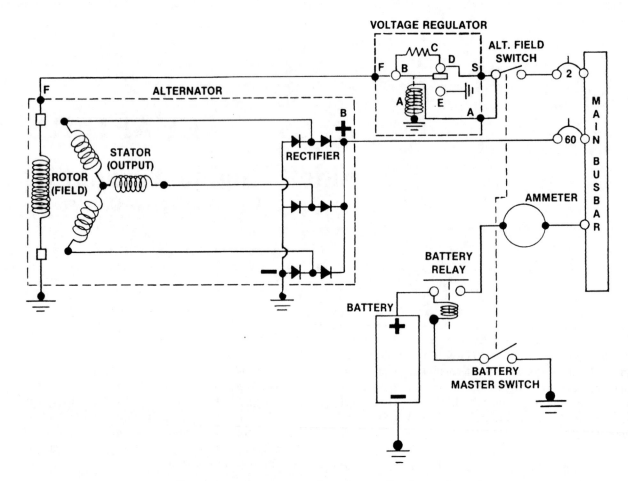

Fig. 1-19 Alternator voltage control schematic

amp circuit breaker. The alternator voltage applied to the bus is also applied to the voltage sensing coil of the regulator through the 2 amp field breaker and alternator field switch.

As the alternator voltage rises above 14 volts, the voltage sensing coil opens contacts **B** and **D**, this forces the field current to go thorugh the resistor **C** and reduces the field excitation. Less field excitation means less alternator output voltage which is sensed by the regulator voltage sensing coil. The contacts **B** and **D** close under 14 volts due to return spring tension on the contacts. This shorts out the resistor **C** and allows full field excitation. The output voltage increases.

This operation is repeated many times per second with the result that the output voltage is maintained at a mean level of 14 volts ±.5 volts. The lower contacts **E** on the regulator are used when a low load, high speed alternator operation allows the alternator voltage to creep up to ap-

proximately 15 volts. At 15 volts, the contacts **B** and **E** are made momentarily as the voltage sensing coil moves the arm **B** down to **E**. This connects a ground to both ends of the rotor field circuit and depresses the output voltage down below the regulated voltage. If the low load, high speed condition persists, the voltage will gradually creep up again and the operation will be repeated.

During engine closedown, the alternator voltage drops. The battery voltage is applied to the **B−** terminal on the alternator but is blocked by the rectifier diodes. It is, therefore, unnecessary to have a reverse current cutout in this system. When the battery master switch is switched "off," the interlock also removes the field excitation by opening the alternator field switch.

NOTE: With this type of system, the alternator is not self-exciting and must receiver initial field excitation from the battery:

32

Review Questions:

1. Name the disadvantages of DC generators and how an automotive style alternator overcomes these disadvantages.

2. Explain the purpose and function of the three main components in an alternator.

3. Explain how the voltage regulator adjusts voltage downward when the alternator voltage rises.

4. How is the alternator field excited?

5. Explain the operation of the regulator under low load, high speed conditions.

6. Why is it not necessary to have a reverse current cutout in this circuit?

CHAPTER 7

Cessna Single Engine
Electrical Power Systems

In Cessna single engine systems, the system configurations are different for the various model years of the aircraft. All systems, however, follow the same general concept.

All Cessna single engine electrical systems are of the single wire negative ground return type and are initially supplied with power from a lead acid battery. The generated power in each system is supplied in the earlier models from a 35 to 50 amp DC generator. In later models, this generated power is supplied as standard by a 60 amp automotive type alternator.

In recent years, the high performance single engine aircraft have had 24V DC systems, while the standard single engine models were equipped with 12V DC systems. With the 1979 model year, all Cessna single engine aircraft were manufactured with 24 volt systems and this is expected to continue with all future models.

A. Battery Locations and Capacity

In single engine aircraft, the battery locations are generally located in the following positions:

150 series — mounted to the starboard firewall
170 series — mounted to the port firewall
180 series — mounted in the aft port tail cone
170 high performance series — mounted in the aft port tail cone
200 series — in the tunnel structure beneath the engine

The capacity of batteries installed in Cessna aircraft follow three general sizes: 150-170 series, 12V 25 ampere hours; 180 series, 12V 33 ampere hours; and 24V aircraft, 24V 18 ampere hours.

B. Battery and Alternator or Generator Master Switches

The battery master switch on Cessna aircraft is generally placed on the left-hand lower instrument panel. In earlier (pre-1964) models, this switch was a double pole on/off switch which was operated by pulling the switch knob out from the panel.

One pole of the switch completes the ground circuit for the battery relay coil. The other pole switches the generator field circuit on or off. The operation of both sets of contacts is simultaneous. From 1964 to 1970, the master switch was a double pole on/off rocker switch again with simultaneous operation of the contacts. The contact functions were one set for battery relay circuit, one set for the alternator field circuit.

From 1971 on the master switch has been a double pole on/off split rocker type which is interlocked so that selection of the battery rocker will allow only the battery circuit to be energized but selection of the alternator rocker energizes both the battery and the alternator field circuit. This is to ensure that the alternator will be excited from the battery whenever the alternator is selected "on" and the alternator will be de-energized when the battery master is switched off. This arrangement, however, does not prevent the battery master switch from being selected to maintain power should the alternator fail.

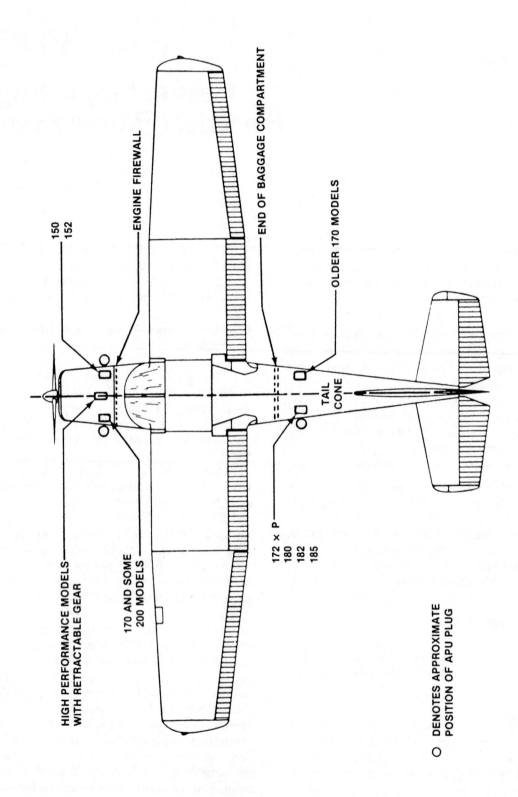

HIGH PERFORMANCE MODELS
WITH RETRACTABLE GEAR

150
152

ENGINE FIREWALL

170 AND SOME
200 MODELS

END OF BAGGAGE COMPARTMENT

OLDER 170 MODELS

172 × P
180
182
185

TAIL
CONE

O DENOTES APPROXIMATE
   POSITION OF APU PLUG

Fig. 1-20 Battery locations for different models of single engine Cessnas.

36

C. Generator Locations (Model to 1963)

The generator on Continental engine aircraft, up to 1963, is mounted in two different positions. On the C150 and C170 series it is mounted on the left upper side of the engine accessory case and is belt driven. To gain access it is necessary to remove the top cowling. On C180, C182, and C185 series the generator is mounted is a similar position to that in a C150 except that access can be obtained through the left-hand cowl door.

The simple system seen in Fig. 1-21 is a early system used on most single engine aircraft up to approximately 1962-63. Its power is supplied by a DC generator. It has a single electrical busbar with no isolation of radio equipment. The external ground power is routed direct from the external power plug to battery contactor output terminal.

D. Alternator Locations

1. 1963 to 1968

The alternators on aircraft with Continental engines, from 1963 to 1968, are mounted in two different positions. On the C150 and C170 series it is mounted on the center rear accessory drive case and is gear driven. On the C180, C182, and C185 series it is shock mounted on the port upper engine and is pulley driven from the accessory drive pulley.

From 1963 to approximately 1968-69 the system was equipped with alternators for generation of power but apart from the main features it remained the same as seen in Fig. 1-21 with a single busbar and no relay controlling ground power availability.

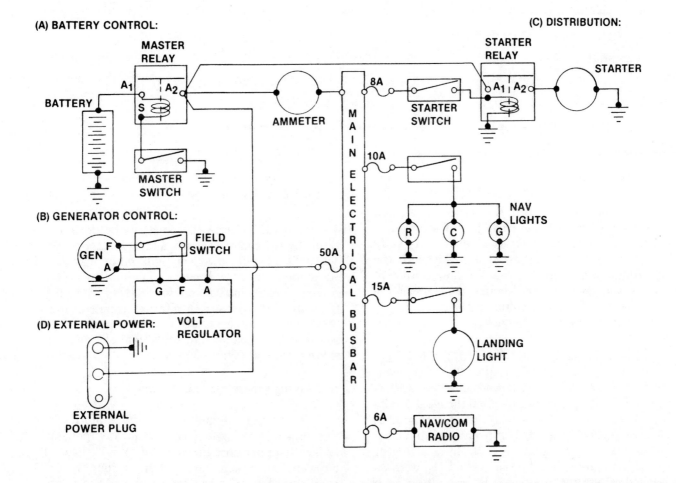

Fig. 1-21   An early generator system

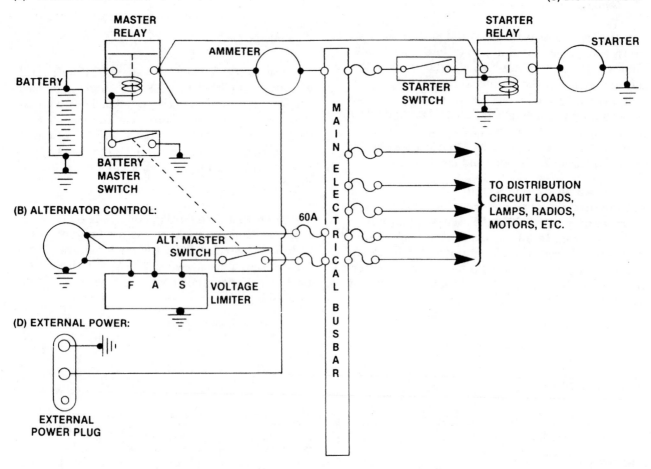

(A) BATTERY CONTROL:

MASTER RELAY

BATTERY

BATTERY MASTER SWITCH

(B) ALTERNATOR CONTROL:

ALT. MASTER SWITCH

F   A   S   VOLTAGE LIMITER

60A

(D) EXTERNAL POWER:

EXTERNAL POWER PLUG

AMMETER

MAIN ELECTRICAL BUSBAR

(C) DISTRIBUTION:

STARTER RELAY

STARTER

STARTER SWITCH

TO DISTRIBUTION CIRCUIT LOADS, LAMPS, RADIOS, MOTORS, ETC.

Fig. 1-22   A typical alternator system from 1963 to approximately 1968-69.

2. 1968 on

In 1968 aircraft in the 170 range were being equipped with Lycoming engines. The positioning of the alternators on these engines is pully driven from the front propeller pulley with the alternator positioned on the port lower front of the engine. On all other aircraft from 1968 on, the positioning of the alternator is the same as discussed in the previous paragraph.

From the period 1968-69 to 1971, the system seen in Fig. 1-23 was developed. The significant changes were: a ground power contactor was introduced with polarity protection diode, D_1, in its coil circuit. A split bus relay was introduced to isolate the avionics busbar from aircraft power when starting or using an APU for starting or servicing. Protection diodes, D_2 and D_3, were installed to prevent interaction between the starter and APU circuits. Diode, D_4, was installed to clip

spike voltages developed across the battery contactor coil as its field collapsed.

In the 1979 alternator system the design is basically the same with only a few changes made. The split bus relay has been replaced by a switched avionics master breaker. The master relay closing circuit introduced for battery charging from the APU. It has an alternator control unit instead of a regular voltage limiter. A low voltage warning light has been installed and all the 1979 systems are 28 volt.

E. Voltage Regulator Locations

The voltage regulators on the majority of Cessna single engine aircraft are located generally on the upper port firewall and may be located adjacent to a radio noise filter in roughly the same location. The regulators of various types differ in outward appearance:

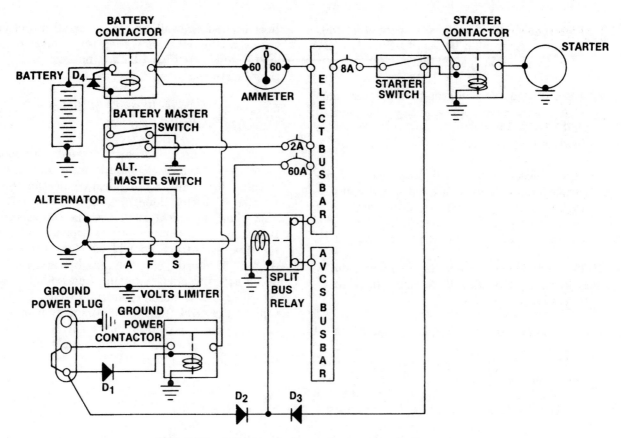

Fig. 1-23   Advanced alternator system design

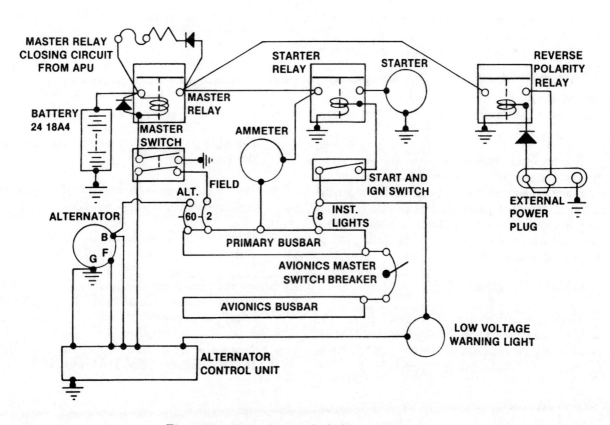

Fig. 1-24   1979 advanced alternator system

1. Electromechanical generator regulator unit: This is a rectangular box approximately 5" × 3" × 2½" and is colored black.

2. Electromechanical alternator regulator: This is a square box approximately 3½" square and is blue in color. It has a cover which is held on by 2 rivets.

3. Hybrid electromechanical, field relay alternator regulator: This is a square box approximately 3½" square and to all intents, looks the same as the electromechanical type. The cover, however, is held on by 2 screws.

NOTE: The above three units are all used on 14 volts systems. On 28 volt systems, fully solid-state regulators are used.

4. 28V solid-state alternator regulator: These regulators are rectangular approximately 5¼" × 3¼" × 2" and are silver in color. They are easily recognized by the fact they have cooling fins or ridges and are connected to the electrical system by moulded rubber plugs.

F. Relays and Contactors

In all single engine Cessnas the main switching functions of battery power, starter supply, and ground power are done by heavy duty electromagnetic relays which are also known as contactors. Modern aircraft will have the following contactors: battery contactor or relay, starter contactor or relay, and ground power contactor or relay.

Earlier light singles will have starter contactors, but use a pull switch which is mounted on the starter operated by a Bowden cable from the pilot's instrument panel. Older aircraft may not have a ground power contactor but will have a direct positive connection from the APU plug to the output of the battery contactor.

1. Contactor locations

General locations for the contractors are: the battery contactor on the battery box or close to it on adjacent structure; the starter contactor will be close to the starter in most cases on the star-

board upper firewall; and the ground power contactor will be close to the battery contactor and APU plug. See Fig. 1-20 for battery and APU plug locations.

G. Actual Aircraft Wiring Routing and Identification

The wiring diagrams covered previously show the system in schematic form only and do not give locations and physical connection, break points, or wiring identifications. Figs. 1-25 to 1-28 show a typical 1970 Cessna 180 electrical power system. This system is broken down into four parts: battery control to electrical busbar (Fig. 1-25); starter and split busbar control (Fig. 1-26); external power starting and split bus system (Fig. 1-27); and the alternator charging system and control (Fig. 1-28). Inspection of these routing diagrams will enable identification of location points, to point wiring details and actual break points in the systems.

Review Questions:

1. Where is the battery located on C170 series aircraft?

2. What is the capacity of the battery in a 1975 C150 aircraft?

3. What is the general location of an APU plug?

4. Explain the operation of the split rocker switch used on post-1971 aircraft.

5. On what series aircraft would you find a gear driven alternator?

6. How would you recognize an electromechanical voltage regulator?

7. Contactor is another name for which components?

8. Explain the main differences between a 1958 aircraft and a 1971 aircraft in terms of electrical power power.

9. Name the significant components connected to the wires identified DA1 and KA2.

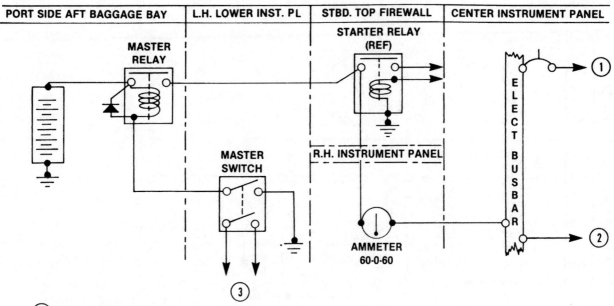

PORT SIDE AFT BAGGAGE BAY	L.H. LOWER INST. PL	STBD. TOP FIREWALL	CENTER INSTRUMENT PANEL

① TO STARTER SYSTEM

② TO BUSBAR SPLITTING SYSTEM

③ TO ALTERNATOR SYSTEM

Fig. 1-25   1970 Cessna 180 battery control to electrical busbar wiring routing chart

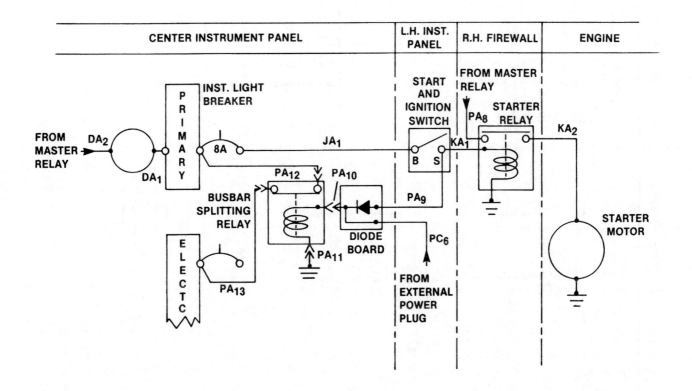

Fig. 1-26   1970 Cessna 180 starter and split busbar control wiring routing diagram

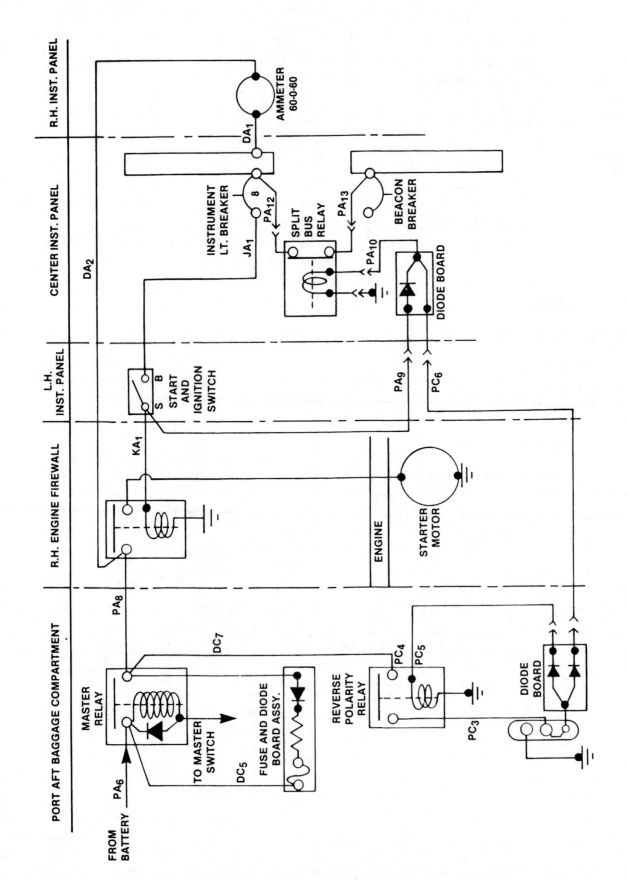

Fig. 1-27 External power, starting and split busbar wiring routing chart for a 1970 Cessna 180

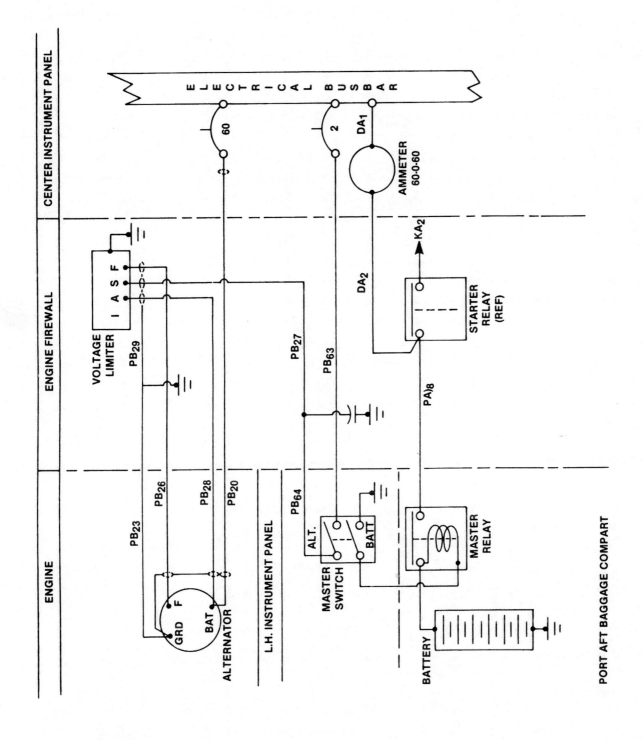

Fig. 1-28 1970 Cessna 180 wiring diagram for the alternator system and control

43

AIRCRAFT ELECTRICAL SYSTEMS

PART II
TWIN ENGINE

CHAPTER 8

Twin Engine
Electrical Systems

The battery supplies the main busbar through a master control and a battery monitoring device. The battery master control must be readily accessible to the pilot.

The generators must be capable of independent operation and of operating in a shared load configuration with controls which will automatically carry out these functions when selected. The generator on/off switches must be readily accessible to the pilot.

A. Operation

Selection of the battery control to "on" will allow power through the battery control, and battery monitor to the main electrical busbar.

Selection of the port generator to "on" will allow generator power to feed the main electrical busbar through the port generator control and generator protection.

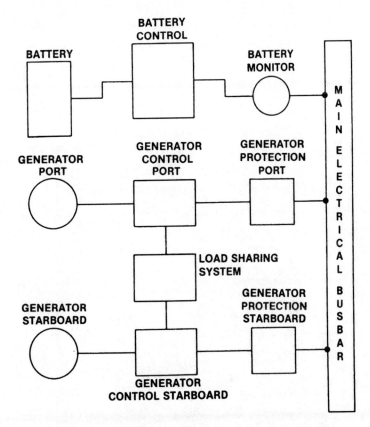

Fig. 2-1 Typical twin engine electrical system

Selection of the starboard generator to "on" will allow starboard generator to feed the main electrical busbar through the starboard generator control and generator protection. When two generators are connected to the busbar the load sharing system ensures that each generator supplies an equal share of the load being drawn from the busbar.

CHAPTER 9

Twin Engine Battery Control Circuit

The battery control circuit on the majority of light to medium twin engine electrical systems resembles the battery control circuit used on the single engine electrical systems. This is the conventional system using a battery relay controlled by a battery master switch. The battery master switch is usually connected in the ground side of the relay coil circuit and when switched on completes the ground circuit for the battery relay coil.

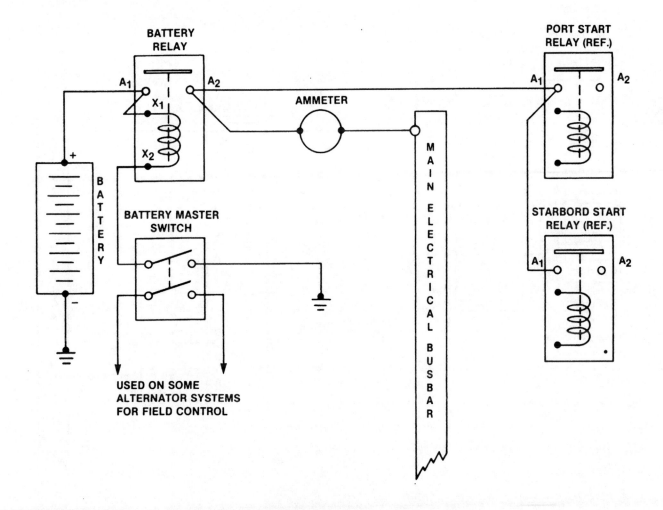

Fig. 2-2   Battery control circuit

A. Operation

Selection of the battery master switch to "on" completes the ground circuit for the battery relay coil. This allows current to flow from the battery positive (+) to A_1 terminal on the relay, then through link to X_1, through the coil through X_2, through the switch to ground and back to battery negative (−).

The relay coil energizes and pulls the contact arm down to bridge contacts A_1 and A_2. Power is applied from the battery through relay contacts A_1 and A_2 to the port and starboard start relay contacts A_1. Power is also applied to the main electrical busbar through the ammeter.

Opening the battery master switch contacts de-energizes the battery relay coil. Spring tension opens the battery relay contacts and removes battery power from the busbar and starter relays.

B. Typical Piper Aztec "C" Battery Control Circuit

Many systems include a provision for using one ammeter instrument to monitor battery charge or discharge and generator output current. Such an arrangement is shown on Fig. 2-3 as used on the Aztec C, but other aircraft use similar arrangements. The principle is to use a shunt in each line to be monitored and to switch the relevant shunt output leads through a multi-position selector switch to a common ammeter.

1. Operation of ammeter selector system

The function of the battery control system is identical to that discussed in chapter 2, with the exception that current from the battery to the busbar is routed through the ammeter shunt.

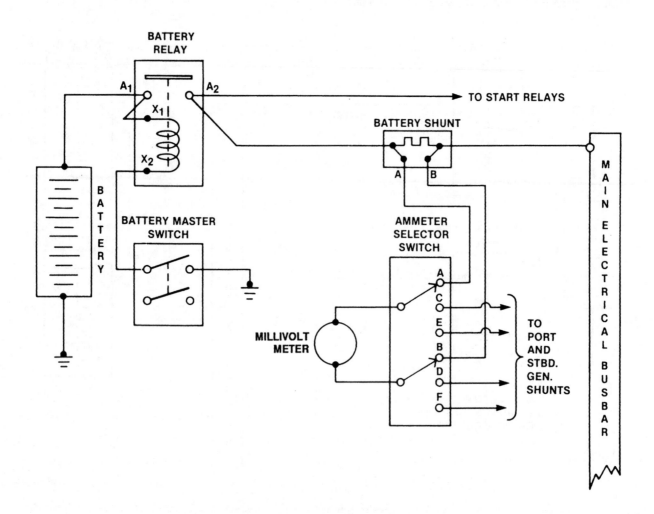

Fig. 2-3 Aztec system with ammeter selector switch

Current flowing through the shunt causes a millivolt drop to be developed across points A and B on the shunt proportional to the current flowing.

With the ammeter selector, as in the diagram, the millivolts developed as applied from the shunt through pin A and B on the selector switch to the millivolt meter and the millivolt meter will read the number of millivolts applied. The scale of the millivolt meter is calibrated in amperes and the instrument indicates the number of amperes flowing which produces the amount of millivolts applied.

Selection of the ammeter selector to C and D will monitor port generator output current, and

selection to E and F will monitor starboard generator output current.

C. Basic External Power System

On twin engine electrical systems, the external power system is similar to that found on single engined aircraft. In its most basic form the external power is applied through an external power plug and heavy gage wire to the output terminal of the battery relay.

Power is available to all circuits when the external power is connected and switched on. This system is employed on most of the earlier generation of twin engined aircraft and many of these aircraft are still operating today.

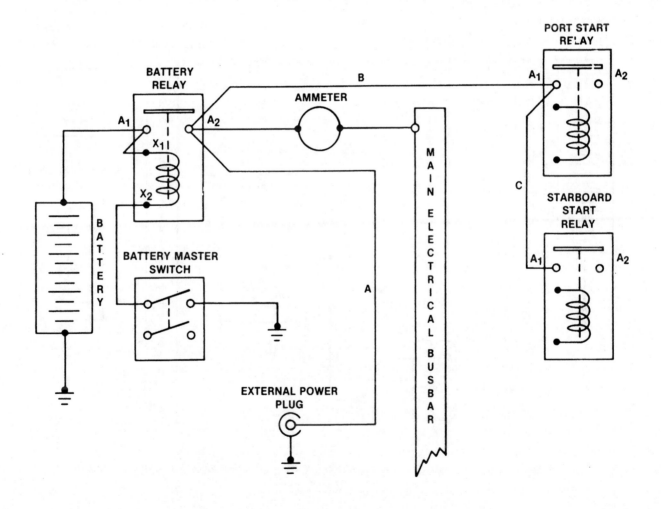

Fig. 2-4 Schematic of a battery control with a basic external power system

1. Operation

The external power is connected to the external power plug. Next the external power is switched "on". Power is applied through wire **A** to terminal A_2 of the battery relay. Power is now available to the busbar through the ammeter and to the port and starboard starter relay inputs A_2 terminals through wires **B** and **C**.

NOTE: The battery relay need not be energized to apply external power to the busbar.

D. Relay Controlled External Power System — Later Model Aircraft

One of the deficiencies of the basic external power system is that there is no protection against a reverse polarity being connected to the external power plug and thus leaving the aircraft circuitry prone to damage.

More recent systems incorporate relay control for the positive line with the relay coil circuit utilizing a diode for polarity protection.

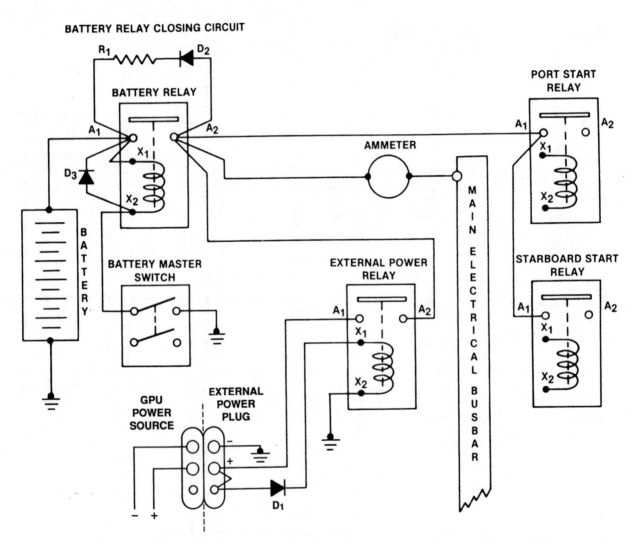

Fig. 2-5   *Advanced battery control with a relay controlled external power circuit*

Also on later model systems, a battery relay closing circuit is added to allow the aircraft battery to be charged in the aircraft.

1. Operation

Power is applied to the external power plug. Provided that the positive is on the centre positive pin current will flow through the link to the small pin, through diode D_1 through X_1 on the relay coil through the relay coil to X_2 and to ground.

The external power relay energizes and closes its contacts. Power is applied from the centre positive pin through contacts A_1 and A_2 of the external power relay to contacts A_2 on the battery master relay. This power is also applied through the ammeter to the aircraft main busbar and to the A_1 contacts on the port and starboard starter relays.

The battery relay closing circuit consists of the diode D_2 and resistor R_1. If the aircraft battery is low, external power is applied through the diode D_2 and resistor R_1 to the battery relay contacts A_1 through the link to X_1 and the coil.

By closing the master switch, current from the external power unit will flow through the coil and energize the battery relay, closing its contacts. Power can now flow from the external power unit through the battery relay and charge the batter.

Diode D_2 prevents the battery from feeding the busbar with the battery relay de-energized. Resistor R_1 drops the external power voltage so that it is just enough to close the battery relay but not to charge a good battery.

Diode D_3 is to clip high voltage spikes across the battery relay coil when it is de-energized by passing the high inductive voltage developed by the coil through the battery to ground. This prevents avionics unit damage.

2. Warning

When preparing to connect external power to any aircraft system, be sure of the following:

1. that the external power unit is compatible with aircraft voltage system;

2. that the external power unit plug is correctly wired as to polarity.

E. Emergency Busbars

On some aircraft systems, provision is made to supply electrical power to the alternator field circuit should the battery relay circuit fail. In other aircraft, power switching is provided direct from the battery to the overhead lights to allow the overhead lights to be selected in order to provide lighting to find the master switch and other controls in the dark. These circuits all take their supply from the input side of the battery relay and must be checked to ensure that they are not left on which would, of course, drain the aircraft battery.

1. Operation

The system operates normally until such time as a failure occurs in the battery relay circuit. This could be open circuit contacts A_1 and A_2 or an open circuit in the coil circuit to ground. Under these conditions, power would not be available to the electrical busbar and both alternators would fail.

Selection of the alternator emergency switch to "on" will feed power direct from the battery through battery relay terminal A_1, through the emergency field circuit breaker and alternator emergency switch to the output side of the left and right field circuit breakers.

The fields of the alternators are re-excited and the alternators will come back on the busbar to supply all loads. It should be noted that the alternators cannot charge the battery as the battery relay contacts are open.

NOTE: The alternate emergency switch has a red guarded handle which when lifted becomes obvious that the system is not in a normal function. This helps to ensure that the emergency system is not left selected and discharging the battery through the alternator fields.

Review Questions:

1. How does the battery master switch energize the battery relay?

2. How is the battery relay opened when switched off?

3. What is the purpose of the battery shunt?

4. What points are important to remember when using an external power source?

5. If diode D_2 on circuit Fig. 2-5 was shorted, what would happen?

6. If the alternator emergency switch was left on what would be its result?

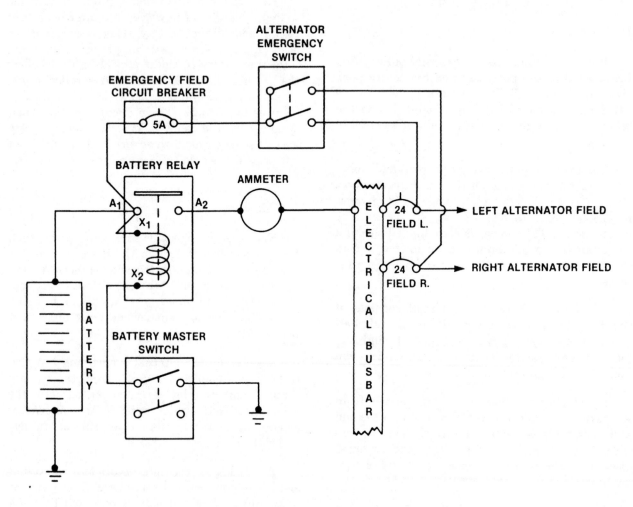

Fig. 2-6 Alternator emergency field excitation

CHAPTER 10

Twin Engine Starter Control Systems

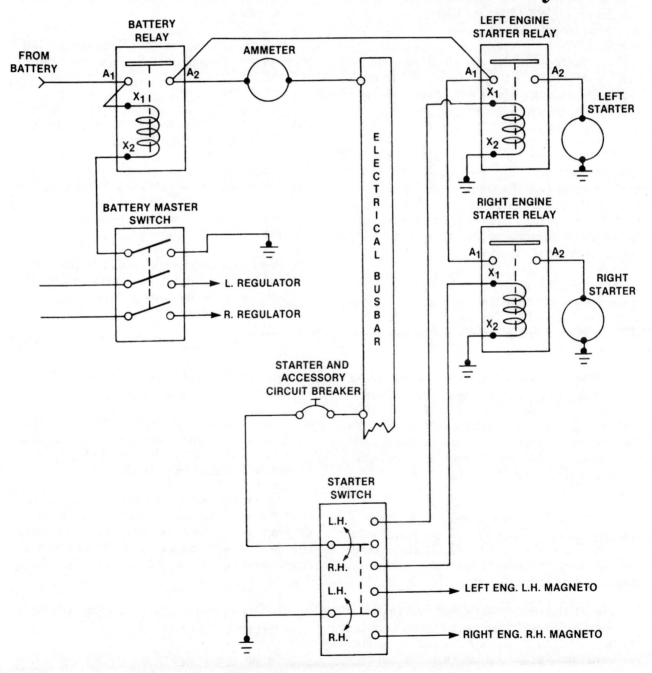

Fig. 2-7 Piper Seneca II starter system

Starter control systems on twin engined aircraft can vary from very simple functional systems to fairly complex fully integrated starter and ancilliary systems such as boosted magneto inputs and fuel priming systems.

Light twin aircraft, such as the Piper Aztec, have a relatively simple system, while the Cessna 421 has a much more complex type of arrangement.

A. Operation

Power is applied from the battery to terminal A_1 on the battery relay.

The battery master switch is selected to "on". Current flows from terminal A_1 through link to X_1, through relay coil to X_2 and through the battery switch to ground. The battery relay energizes, closing contacts A_1 and A_2.

Battery power is applied through the battery relay contacts to the A_1 terminals on both starter relays and through the ammeter to the electrical busbar. Power on the busbar is available at the starter and accessory circuit breaker and through to the center terminal of the top portion of the starter switch.

Selecting starter switch to left hand position applies power to the left engine starter relay coil energizing the relay. The relay contacts close allowing power through to the left starter motor turning the engine.

During this operation, the lower portion of the starter switch is also moved to the left hand position, causing the left engines right hand magneto to be grounded through the switch. This kills the right hand magneto and only allows the left hand magneto which has the impulse coupling fitted to provide a correctly retarded spark for starting.

When the engine starts, the starter switch is released and it springs back to its centre positions de-energizing the left engine starter relay and removing the ground from the right hand magneto.

The starter motor stops and the engine is now running on both magnetos. Operation of the starter switch to the right hand position causes the same series of events, but this time from the right engine.

1. Operation of twin engine starter system with booster vibrator

With battery relay energized, contacts A_1 and A_2 close allowing battery power through the ammeter **A** to the electrical busbar and to terminals A_1 on the left and right starter relays **B**.

Power from the busbar is applied through the starter circuit breaker **C** to C_1 terminal on the right starter push switch **D** along the interconnecting wire to the C_1 on the left hand starter push switch and up to the "in" terminal on the booster vibrator **E**.

Left engine mag switches **F** selected to "on" removes the ground from the magneto primary coils.

Pushing the left starter push switch **D** changes its 4 contacts over to the normally open (N.O.) position with the following results.

1. C_1 contact connects battery power to the left starter relay coil X_1 terminal **B** and energizes the starter relay. This allows battery power from the battery relay A_2 terminal to left starter motor **J**.

2. Contact C_2 connects a ground to the switch lead on the right hand magneto primary circuit **G** and disables the right hand magneto **G** on the left engine.

3. Contact C_3 connects booster vibrator output **E** from the "BO" terminal on the booster vibrator **E** to the retard terminal on the left engine left hand magneto **H**.

4. Contact C_4 connects booster vibrator output **E** from the "BO" terminal on the booster vibrator **E** to the switch terminal on the left engine left hand magneto **H**.

The engine turns and the boosted inputs allow the left hand magneto **H** to fire and start the engine. Releasing the left hand starter push switch **D** allows contact C_1 to C_4 inclusive to return to their original position with the following effects:

1. The left hand starter relay **B** de-energizes and stops the starter motor **J**.

2. The left engine right hand magneto **G** is live as the ground has been removed.

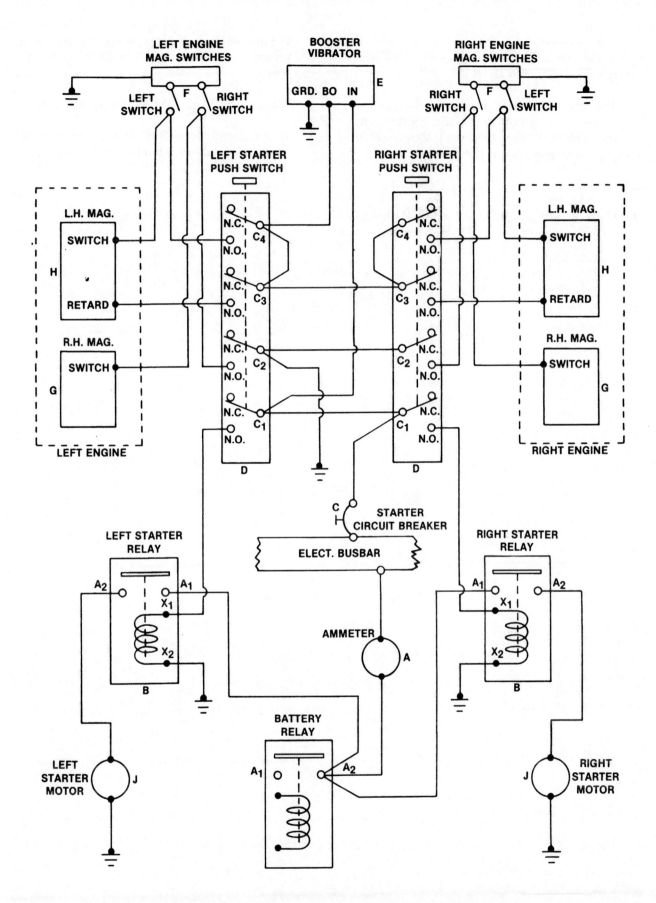

Fig. 2-8   Twin engine starter system with booster vibrator

3. The boosted output is removed from the switch and retard terminals of the left engine left hand magneto **H**.

The left engine now runs normally on both magnetos. The operation of the right engine starter system as that described above, except that the right hand components are used.

Review Questions:

1. With the left starter switch selected on what indication whould show on the ammeter?

2. During the starting operation one magneto primary is automatically grounded. Explain why.

3. What is the purpose of a booster vibrator?

4. If the circuit breaker C in Fig. 2-8 fails to open, what effect would this have on the starter circuit?

CHAPTER 11
Twin Engine Generator Control and Load Sharing

A. *Electromechanical Regulator*

In twin generator electrical systems, each engine drives a generator and its output is connected to a common busbar. As the engine speed varies on each generator, one generator may produce more output and take more than its share of the total load.

In aircraft with DC generators, it is necessary to have a method of automatically adjusting the generator outputs so that each generator carries an equal amount of the total electrical load.

When the generators are controlled by the electromechanical type regulators they are fitted with an extra winding on the voltage regulator unit to sense a rise in that particular generators output voltage.

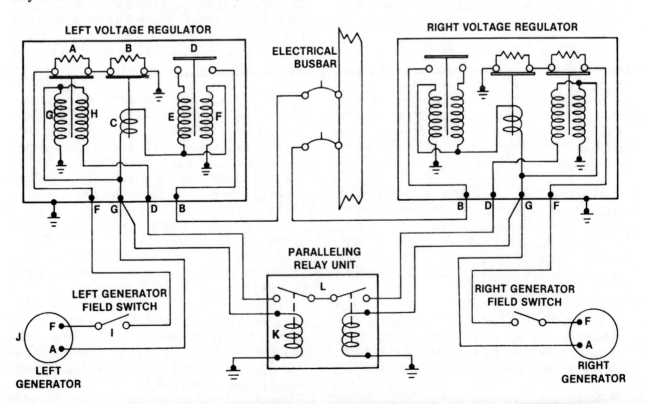

Fig. 2-9 An electromechanical regulator for a twin engine generator system

59

The extra windings are connected to each other through a paralleling relay so that the effect of a rise in voltage on one generator will produce the same effect on the other generator. The paralleling relay has two coils which are connected to their respective generators output and energized when their respective generator is at normal regulated voltage. Each coil closes contacts which are in series with each other and complete the paralleling circuit between the generators.

1. Operation

a. Bringing generator onto the electrical busbar

Battery master swith "on". Left engine started. Left generator field switch **I** placed "on".

Generator **J** builds up its voltage from residual magnetism in its pole shoes. Field current flows from generator terminal "A" through the field switch **I**, through the "F" terminal on the left voltage regulator unit, through contacts **A** and **B** to ground. This causes increased excitation and voltage builds up rapidly to approximately 12.6 volts. This 12.6 volts is applied from the "A" terminal on the generator through the "G" terminal on the voltage regulator unit and through coil **C** to coil **F**. The 12.6 volts applied across coil **F** causes contacts **D** to close and connects generator output voltage to the busbar.

The voltage output of the generator continues to rise up to 14 volts. As soon as voltage output exceeds 14 volts, this voltage applied across coil **G** causes contacts **A** to open. Contacts **A** opening causes the resistor across the contacts to be inserted in the field circuit and reduces the excitation which reduces generator output below 14 volts.

As soon as the voltage drops below 14 volts, spring pressure on the contacts **A** returns them to the closed condition and increases field excitation raising the voltage up to 14 volts and above. This process is repeated many times per second and keeps the voltage relatively constant at a mean level of 14 volts ± .5 volts.

The generator is connected to the busbar charging the battery and current through coil **E** and contacts **D** to the busbar. The effect of coil **E** is to assist coil **F** in keeping contacts **D** closed.

Generator output from terminal "G" on the regulator unit is applied to coil **K** on the parallel-ing relay unit which energizes and closes contacts **L**.

The procedure for bringing the right hand generator onto the busbar is identical to that described in the above paragraphs.

2. Paralleling Operation

With both generators connected to the busbar, both contacts **L** in the paralleling relay unit are closed. This connects the paralleling coils **H** in both regulators together through the "D" terminals on the regulators.

If the voltage rises on the left generator, the voltage at the top of coil **H** in the left regulator will be higher than the voltage at the top of coil **H** in the right regulator. Current will flow through coil **H** on the left regulator through the paralleling relay contacts **L** and through coil **H** on the right regulator due to this small difference in voltage.

This current will cause a reduction in the left generator voltage as it is flowing through coil **H** from the top in the left regulator and an increase in the right generator voltage as it is flowing through coil **H** from the bottom in the right regulator. The current through coil **H** due to a voltage imbalance assists or opposes the magnetic effects of the voltage coil **G** depending on which way the current flows through the paralleling circuit.

Current control on a twin generator system is the same as a single generator system. Coil **C** carries full generator output current and if the rated ouput is exceeded, contacts **B** open. This inserts the resistance connected across the contacts into the field circuit causing a reduction in voltage which in turn reduces output current.

As the paralleling relay coils **K** are connected to their particular generator outputs failure of either generator will automatically disconnect the paralleling circuit by opening one of the series contacts **L**.

This system has conventional reverse current cut out protection where a flow of current from the battery and busbar through contacts **D** and coil **E** will cause contacts **D** to open and disconnect the generator from the battery and busbar whenever a generator voltage is lower than the battery voltage.

NOTE: It is important that when setting up the voltage regulators on twin generator system that the voltage settings are precisely the same for similar loads.

B. Carbon Pile Voltage Regulator Principles

In generator systems with an output of 50 amps or more, it has been the convention to use not an electromechanical contact type regulator unit, but to use what is known as a carbon pile regulator.

The carbon pile regulator uses an electromagnet to vary the pressure on a stack of carbon discs (carbon pile) and so vary the resistance of the stack. This stack or pile of carbon discs is inserted in series with the field of the generator so that variations in pile pressure will vary the resistance and consequently vary the output of the generator by varying field excitation current.

The changes in resistance due to changes in output voltage of a generator system fitted with a carbon pile regulator are smoother than with an electromechanical contact type regulator.

A carbon pile regulator consists basically of the following parts:

1. Magnet core assembly on which is mounted a magnet coil in the form of a circular winding.

2. The magnet core attracts a soft iron armature which is mounted on leaf springs which oppose the magnet.

3. A carbon pile which is contained in a ceramic tube and held in place by a carbon insert attached to the armature and a pile compression screw with carbon insert in the other end frame of the assembly.

4. A bimetallic ring which is placed under the leaf springs on the armature assembly and distorts so as to correct for temperature increases in the regulator under working conditions.

5. The whole appearance of the regulator is a cylinderical finned unit approximately 3-1/2" in diameter and 6" in length.

6. Majority of carbon pile regulators in North America are supplied with plug-in connec-

tions and can be easily replaced when servicing the system.

1. Characteristics of carbon piles

a. Carbon has a negative coefficient of resistance which means that as temperature increases, resistance decreases.

b. By using many carbon washers in a pile, we create many junctions which are of a relatively rough surface texture.

c. More pressure on the pile will cause more carbon to be contacting between the washers and decrease pile resistance.

d. Less pressure on the pile will cause less surface to contact and increase pile resistance.

e. When the regulator is not working, the greatest pressure is exerted on the pile which causes minimum resistance in the field circuit.

2. Operation

When generator armature turns, the voltage output builds up rapidly through residual magnetism in the field system.

Field current flows through from G− through the shunt field winding out of pin S on the generator, through the fully closed carbon pile to the G+ terminal on the regulator.

When the output voltage builds up to approximately 28 volts, the magnet coil has sufficient magnetic pull to cause the armature to reduce pressure on the pile. This opens the carbon pile and increases the resistance in the field circuit which reduces field excitation and reduces voltage output.

As the voltage drops below 28 volts, the leaf spring tension overcomes the electromagnetic attraction and the carbon pile closes. This reduces field resistance, increases field excitation and increases output voltage.

The voltage regulator is set up so that electromagnetic pull and spring tension equal each other at around the 28 volt level so that the regulator keeps the voltage at a mean level of about 28 volts.

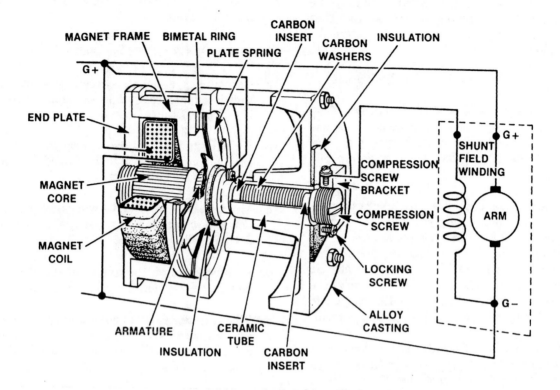

Fig. 2-10 Construction of a simplified carbon pile regulator circuit

All carbon pile voltage regulators work on the above principle, but regulators used on twin engine systems have an extra coil wound on the same former as the voltage regulator which is used when paralleling the generators to the electrical busbar. This coil is known as an equalizing coil.

C. *Generator Control and Load Equalizing (Carbon Pile)*

On aircraft which are fitted with 50 amp or higher generators, it is a common practice to use carbon pile voltage regulators. When the generators are connected in parallel to the busbar, it is necessary to provide control so that both generators share the total load equally. This is achieved by the use of an equalizing circuit which automatically adjusts each regulator as their outputs vary.

The equalizing system has a low value resistance connected in each of the generator ground leads which develops a voltage in proportion to each generators particular current output. By

connecting the two regulator equalizing coils in series with the tops of the two resistors, the system senses and adjusts the generator outputs so that they are in balance.

In this diagram the identification of the components are listed as a carbon pile in series, a voltage trimmer potentiometer, a voltage sensing coil and a equalizing coil.

1. *Operation of a simple equalizing system*

Each generator is run up to normal engine speed and connected through the reverse current relays **N** and **O** to the battery busbar **U** and through the closed battery relay contacts **T**, to charge the battery **X**. If each generator system is producing precisely the same output voltage of say, 28 volts, then the current delivered by each generator to the battery busbar **T** and aircraft loads will be the same; ie. if the total current loads on the battery busbar is 80 amps and we have two 50 amp generators supplying the loads, each generator should supply 40 amps to the battery busbar **T**.

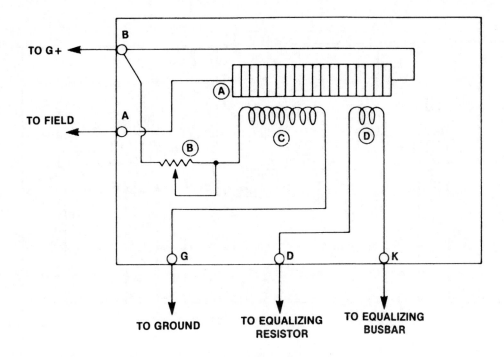

TO G+

TO FIELD

B

A

C

D

G — TO GROUND

D — TO EQUALIZING RESISTOR

K — TO EQUALIZING BUSBAR

A CARBON PILE IN SERIES WITH FIELD CURRENT

B VOLTAGE TRIMMER POTENTIOMETER

C VOLTAGE SENSING COIL CONNECTED ACROSS GENERATOR OUTPUT

D EQUALIZING COIL ACTS WITH OR AGAINST VOLTAGE COIL DEPENDING ON WHICH WAY CURRENT FLOWS THROUGH IT

Fig. 2-11   A circuit diagram of a carbon pile regulator with an equalizing coil

If each generator is supplying 40 amps, then 40 amps must be flowing through each of the equalizing resistors **A** and **B**. The value of each equalizing resistor is such that at full generator output of 50 A, a voltage of .5 volts will be present between the top of the equalizing resistor and ground. This means that the value of each equalizing resistor equals $E \div I = R = .5 \div 50 = .01$ ohms.

The equalizing circuit connects the top end of each equalizing resistor together after going through the left equalizing coil **P**, the left equalizing switch **R**, the right equalizing switch **S**, and the right equalizing coil **Q**.

With equalizing switches **R** and **S** closed and assuming conditions of 40 amps being supplied by each generator, the voltage at the top of each of the equalizing resistors **A** and **B** will be 40 × .01 = .4 volts each. As there is no difference in the voltage between the top ends of the equalizing resistors, no current will flow through the equalizing circuit. The circuit under these conditions is said to be balanced.

If we assume that these two generator systems run for a total time of 50 hours, variations in generator output voltage will occur and the generator output on one generator may become higher than that on the other due to:

(a) dissimilar wear rate on the carbon piles
(b) mechanical dissimilarities in the regulator armature
(c) general deterioration on contacts in the regulator

The result will be that one regulator will control its generator at, say 28 volts, and the other

63

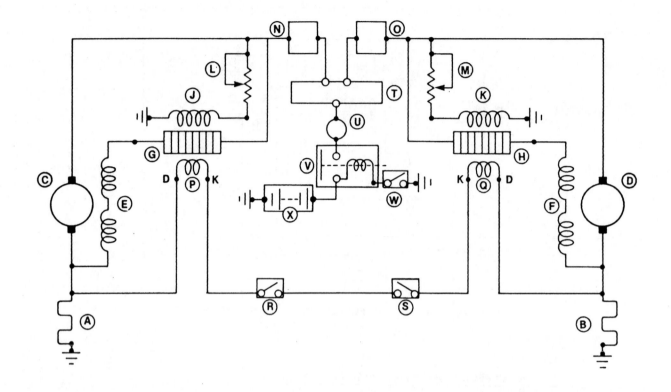

A   EQUALIZING RESISTOR LEFT GENERATOR

B   EQUALIZING RESISTOR RIGHT GENERATOR

C   LEFT GENERATOR ARMATURE

D   RIGHT GENERATOR ARMATURE

E   LEFT GENERATOR FIELD

F   RIGHT GENERATOR FIELD

G   LEFT REGULATOR CARBON PILE

H   RIGHT GENERATOR CARBON PILE

J   LEFT REGULATOR VOLTAGE COIL

K   RIGHT REGULATOR VOLTAGE COIL

L   LEFT VOLTAGE TRIMMER RESISTOR

M   RIGHT VOLTAGE TRIMMER RESISTOR

N   LEFT REVERSE CURRENT RELAY

O   RIGHT REVERSE CURRENT RELAY

P   LEFT REGULATOR EQUALIZING COIL

Q   RIGHT REGULATOR EQUALIZING COIL

R   LEFT EQUALIZING SWITCH

S   RIGHT EQUALIZING SWITCH

T   BATTER BUSBAR

U   AMMETER

V   BATTERY RELAY

W   BATTERY SWITCH

X   BATTERY

Fig. 2-12 A simple equalizing system schematic

regulator will control its generator at 27.5 volts, due to excessive wear on the lower voltage regulator.

Under these conditions with the left regulator controlling its generator at 27.5 volts and the right regulator controlling its generator at 28 volts, the voltages developed at equalizing resistors **A** and **B** changes. The 27.5 voltage generator will supply 20 amps to the busbar and the 28 voltage generator will supply 60 amps to the busbar. The voltage developed at equalizing resistor **A** will be $20 \times .01 = .2$ volts and the voltage at equalizing resistor **B** will be $60 \times .01 = .6$ volts.

We have equalizing resistor **B** delivering .6 volts to the equalizing circuit and equalizing resistor **A** delivering .2 volts. This difference of .4 volts causes current to flow from equalizing resistor **B** through equalizing coil **Q** from D to K, through equalizing switches **S** and **R**, through equalizing coil **P** from K to D, to equalizing resistor **A** and to ground.

The equalizing coils interact with the regulator voltage coils **K** and **J** as follows:

(a) current through **Q** from D to K opens the right carbon pile and reduces the right generator voltage;
(b) current through **P** from K to D closes the left carbon pile and increases the left generator voltage;
(c) the voltages now balance out and each generator supplies 40 amps to the busbar loads.

If the left generator were higher in voltage than the right generator, the opposite effects would be produced in the equalizing coils with the same results.

NOTE: The equalizing circuit can only counter-act small discrepancies in voltage between generators and will only work successfully with less than a .5 volt difference between generators. It is important therefore, that care is exercised when carrying out a generator balance that precise settings are made to each of the voltage regulators.

D. Generator Balancing and Adjustment

Whenever it becomes necessary to adjust the voltage regulators on the system, the procedures detailed in the aircraft manufacturers manual should be followed.

Always use the most accurate voltmeter available for regulator setting and adjustments. Set both regulators to precisely the same voltage. If you cannot obtain exactly 28 volts, try to come as close as possible on the first regulator and then aim to achieve precisely the same setting on the second regulator.

1. Operation of Cessna 310 generator system

Turn the battery master switch "on", and then the engines are started. Next, turn the left generator switch to "on" and this completes the following functions:

(a) generator field excited and generator voltage increases until left reverse current relay energizes by the generator voltage going through the link from generator to switch terminal on the reverse current relay. Generator is switched onto busbar and is regulated by the left regulator.
(b) the other contact in the generator switch closes and prepares the equalizing circuit for use.

The right generator switch is then placed "on" and this system operates identically to that described above.

The second contact in the right generator switch closes and completes the equalizing circuit from the right system across the wire link to the left system. The equalizing system is now connected in circuit and operates normally.

Review Questions:

1. If contacts L in the paralleling relay open circuited, how would the system react?

2. Why is it necessary to apply at least 12.6 volts across the coil **F** in the voltage regulator unit?

3. What is the purpose of the bimetallic ring in a carbon pile regulator?

4. How is the tension applied to the carbon pile?

5. As the temperature of the regulator increases, what happens to the carbon pile's resistance?

6. When the generated voltage decreases below 28 volts, what happens to the pressure on the carbon pile?

7. Where is the equalizing coil on a carbon pile regulator wound?

8. What are the resistance characteristics of the equalizing coil?

9. A 50 amp twin generator system has two equalizing resistors in its ground circuits,

what voltage would be developed across each resistor at full load?

10. Why is it necessary to carry out an equalizing and balancing adjustment on carbon pile regulator systems?

11. At approximately what difference in voltage between generator will the equalizing circuit be ineffective?

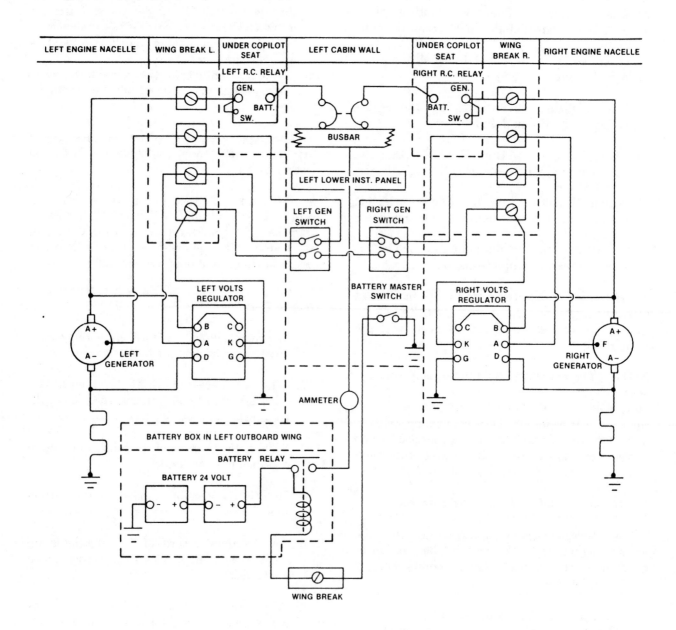

Fig. 2-13   Wiring diagram for a typical twin generator equalized circuit on a Cessna 310

Any DC generator system requires built in protection to prevent the battery from feeding excess current through the armature of the generator when the generator voltage is less than that of the battery. The most obvious time for this to occur is during engine shut down. The method of overcoming this problem is to use a relay which is sensitive to a current flow from the battery to the generator armature.

Such a device is a reverse current relay or cut out which is an inherant part of any DC generator control circuit.

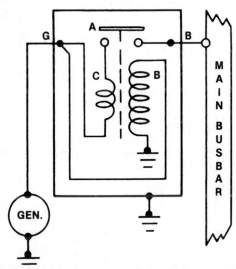

A A PAIR OF SPRING-LOADED TO OFF CONTACTS

B A VOLTAGE SENSING COIL TO CLOSE THE CONTACTS

C A SERIES CURRENT SENSING COIL WOUND ON THE SAME COIL FORMER

Fig. 2-14 Reverse current relay

A. Design Considerations

The reverse current relay should automatically disconnect the generator from the battery when the battery voltage is higher than generator voltage. It should also automatically connect the generator to the battery when the generator voltage exceeds battery voltage.

1. Operation of a reverse current relay

The generator produces output and output rises to 25.2 volts. At above 25.2 volts the voltage coil **B** has sufficient electromagnet force to pull the contacts **A** closed against the spring tension. When contacts **A** close, current from the generator flows through the G terminal, C coil and A contacts, to the main busbar and battery. The current through coil C assists the voltage coil in keeping the contacts A closed, due to the direction of current flow through its windings.

As the engine closes down, the generator voltage falls below that of the battery. Current flows from the battery through the main busbar, through contacts A on the reverse current relay and continues through coil C to the generator. As the generator voltage decreases further, this reverse current increases. The magnetic effect of coil C increases but is now opposite to that in the paragraph above. When the magnetic effects of coils **B** and **C** are equal and opposite, spring tension opens the contacts A, disconnecting the generator from the busbar and battery.

B. Switched Reverse Current Relay Circuit

Some aircraft have a slightly modified reverse current relay which enables the flight crew to switch the generator onto the busbar, but retains the function of a reverse current relay. This type of device has an extra voltage coil and a pair of contacts which complete the switch selection circuit to ground.

1. Operation

The generator producing sufficient voltage is 25.2 or over. Voltage coil A energizes and closes pilot contacts B. This prepares the ground circuit for the contactor coil C so that when the generator switch is closed, the contactor coil is energized from A+ through the generator switch, the "SW" terminal, contactor coil C pilot contacts B to ground. This closed the auxiliary contacts D and the main contacts E.

Current from the generator flows through the generator terminal, to the current coil F and the contacts E and D, to the battery busbar. The magnetic effect of the current through the current coil F assists in keeping the pilot contacts B closed.

As the engine slows down, the generator voltage decreases. When the battery voltage is higher than the generator voltage at the battery teminal, current flows through contacts D and E and the current coil F to the generator armature.

As the current through the current coil F is now flowing in the opposite direction, the magnetic effect of the current coil F cancels out the effect of the voltage coil A and the pilot contacts B spring open. This disconnects the ground from the contactor coil C and allows the contacts D and E to open, disconnecting the generator from the battery.

Before the next engine start, the generator switch should be placed to "off".

C. Differential Reverse Current Relays

The later models of aircraft using DC generators invariably use a different type reverse current relay which is known as a differential reverse current relay. This relay operates on the principle that for correct charging there must be a differential in voltage between the generator and the battery busbar. When this voltage differential is attained, the generator is connected to the battery

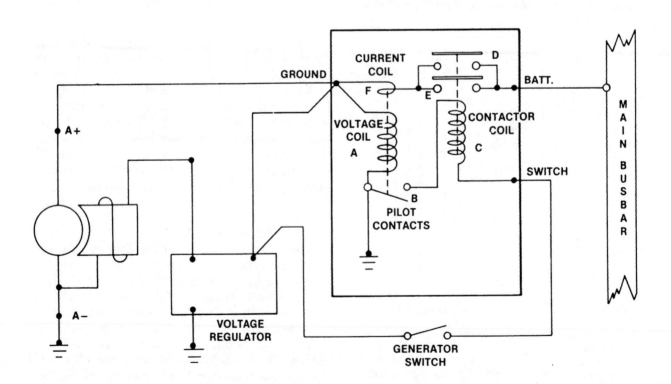

Fig. 2-15 Switched reverse current relay circuit

busbar. The voltage of the generator should be between .35 to .65 volts higher than the battery busbar voltage to allow the generator to be connected to the battery busbar.

1. Operation

The generator builds up voltage and this voltage is applied to the voltage coil **A** through the generator switch and "SW" terminal on the reverse current relay unit. When sufficient voltage has built up, the voltage coil **A** closes its contacts **B**. This allows generator voltage to be applied to the differential coil **C**. When the generator voltage is .35 to .65 volts higher than the voltage on the battery busbar, sufficient current flows through the differential coil **C** to cause it to close the differential relay contacts **D**.

This allows generator voltage from the "SW" terminal to be applied through contacts **D** and across the main contactor coil **E**. The main contactor contacts **F** close and connect the generator to the busbar through the current coil **G**. Current coil **G** assists the differential coil **C** in keeping differential coil contacts **D** closed while current is

traveling from the generator to the battery busbar.

When the generator runs down in speed and voltage, a reverse current flows from the battery busbar through the main contacts **F** and through the current coil **G** in the opposite direction to the generator armature.

When this reverse current builds up to a flow of 10 to 20 amps current coil **G** cancels out the magnetic effect of the differential coil **C**. Contacts **D** open due to return spring tension. This removes the voltage from the main contactor coil **E** and the main contacts **F** open disconnecting the generator from the battery busbar.

The generator voltage decreases further and the voltage coil **A** can no longer hold in its contacts **B**. This disconnects the differential coil **C** from its generator source and also prevents voltage and current feedback from the battery busbar to the generator when the generator is stationary. After engine is closed down, the generator switch is opened to completely disable the system.

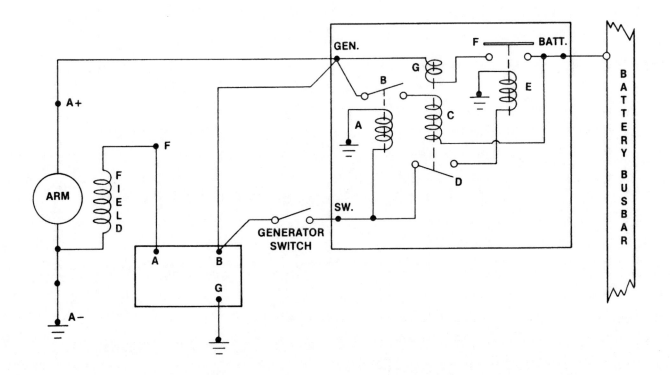

Fig. 2-16 A schematic for a differential reverse current relay system

Review Questions:

1. What are the basic requirements of a reverse current relay?

2. What controls the current through the contactor coil on a switched reverse current relay system?

3. In a differential reverse current relay system what differential in voltage closes the main contacts?

4. At approximately what reverse current will the main contacts open?

CHAPTER 13

Twin Engine Alternator Systems Using Main and Auxiliary Regulators

In the first twin engine systems using alternators, the output of both alternators was controlled by a common voltage regulator. This regulator controlled the field current of both alternators so as to keep the main busbar voltage relatively constant.

The system contains two regulators which can be selected as the main or auxiliary regulator so as to provide a back-up regulation system should one regulator fail.

A. Operation

With the battery master switch "on", battery power is applied to the main electrical busbar **A**. Turn regulator selector switch **D** to main position. Make sure alternator master switch is closed. Power is applied from main regulator breaker **E** through regulator selector switch contacts 10 and 3 to main voltage regulator **J** terminal **B**, then through the internal regulator circuitry and terminal F, through pins 4 and 12 of the regulator

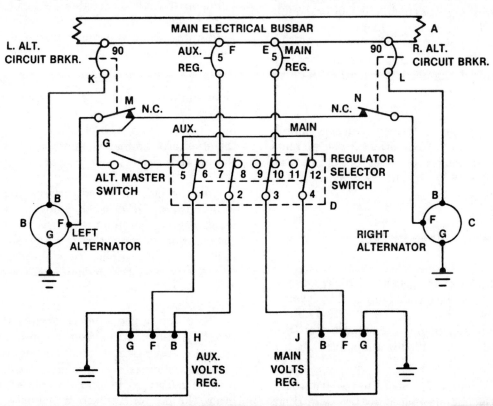

Fig. 2-17 Selectable regulator system

71

selector switch **D**, through the link from pin 12 to pin 5, through the alternator master switch **G** to the common connection between the two microswitches **M** and **N**.

The power is now applied to the F terminals on both the alternators **B** and **C** through the microswitches **M** and **N**. Current flows through the field winding on the alternators to ground. With both engines running the alternators **B** and **C** excite and produce power to the busbar **A** through circuit breakers **K** and **L**.

As the busbar voltage varies the effect is sensed at terminal B of the main voltage regulator **J** which adjusts the field current to both alternators through the path described above.

Should the main regulator **J** fail, selection of the regulator selector switch **D** to the AUX position will substitute the auxiliary volts regulator **H** in to the circuit and the circuit will operate as before using the AUX circuit breaker **F**, pins 7 and 2 on the regulator selector switch, the auxiliary regulator pins 1 and 5 on the regulator selector switch to the alternator master switch then to the alternator fields as described above.

If one alternator fails, it is isolated from the circuit by tripping the main alternator circuit breaker **K** or **L** which are mechanically interlocked to microswitches **M** and **N**. When the circuit breakers are tripped the associated microswitch is opened. This disconnects power from the field of the rejected alternator.

NOTE: The system described above is simplified for explanation of operation. Most systems also include alternator overvolt protection and the capability to monitor alternator current and voltage outputs.

B. *Typical Twin Engine Electrical Alternator System*

Fig. 2-18 shows the schematic diagram for a Piper Aztec fitted with a shared regulator system overvolt relays and an ammeter selection circuit. It should be noted that the system differs from that in Fig. 2-17 on the following points:

1. Ammeter shunts **H**, **I** and **J** have been added to monitor left alternator, battery and right alternator currents.

2. Ammeter selector switch **P** and the ammeter have been added to select and read the ammeter shunt outputs.

3. Overvolt relays **V** and **W** have been connected in the power line to the respective voltages regulators to monitor voltage and open circuit the power to the fields if voltage becomes excessive.

4. R.F.I. filters **K**, **M**, **Y** and **Z** have been installed to reduce radio frequency interference generated by the alternators and regulators.

5. The battery and its control circuitry are shown in Fig. 2-18.

1. Operation

Selection of the master switch **N** on "on" applies battery power to the busbar **A** through the battery relay contacts. This power is also applied to both generator fields through the auxiliary circuit breaker **E**, the regulator selector switch **S** contacts 9 and 3, through the auxiliary overvolt relay **V** pins 1 and 2, through the auxiliary voltage regulator **U** pins B and F, through the regulator selector switch **S** pins 4 and 11, through the master switch **N** pins 4 and 3 to be the common point on the left alternator field microswitch **F** then in parallel through the respective alternator field microswitches to the F terminals on the alternators **Q** and **R**.

The alternators will excite and with the engines running, the alternator outputs will be controlled by the auxiliary voltage regulator. Current flowing from each alternator will go through the alternator ammeter shunts to the busbar and through the battery shunt to charge the battery.

As the current flows through the shunts, a millivoltage drop is developed across the resistance of the shunt which is proportional to the current flowing through the shunt. These millivoltages are applied from the shunts to pins 3, 4, 5, 6, 7, and 8 of the ammeter selector switch **P**.

Selection of the ammeter switch **P** to the appropriate pair of wires from a specific shunt will apply the millivoltage across the ammeter instrument **X** which is really a millivoltmeter whose scale is calibrated in amps. As shown in the diagram, the pair of leads from the battery shunt **I** are connected across the ammeter **X**.

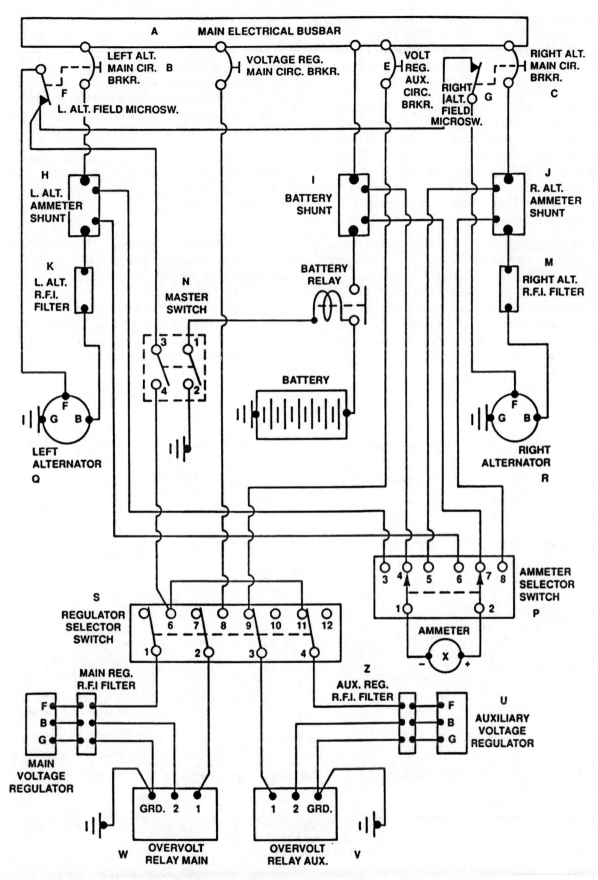

Fig. 2-18   Piper Aztec alternator control system schematic

Selection of pins 3 and 6 on the ammeter selector switch **P** will allow reading of the left alternator output current while the selection of pins 8 and 5 will allow reading of the right alternator output current.

The overvoltage relays **V** and **W** sense the relevant busbar voltage through the regulator selector switch at pins 1 and ground. If busbar voltage should rise above a predetermined level, the overvolt relay senses the rise and opens the circuit between pins 1 and 2. This disconnects excitation power from the alternator field circuits, and shuts down both alternators.

If an alternator should fail, the appropriate alternator main circuit breakers **B** or **C** must be tripped by the pilot. The alternator field microswitches **F** and **G** are linked mechanically to the circuit breakers and the microswitches will open when the circuit breakers are tripped. Tripping the appropriate breakers also isolates the defective alternator field from the field power supply.

C. Alternator Control System Individual Regulators

Later model twin engined aircraft have alternator control systems which have an individual solid state regulator controlling each alternator. The systems also have an extra terminal on the regulator which is a paralleling connection.

This connection is used to sense both alternator field voltages and automatically bring up the output of the lower voltage alternator to that of the highest alternator.

The regulators are completely transistorized and have an adjustment potentiometer to set up the operating voltage of each alternator.

The paralleling function is carried out by special transistors in each regulator which sense a discrepancy in field voltage and apply the necessary correction to balance the system.

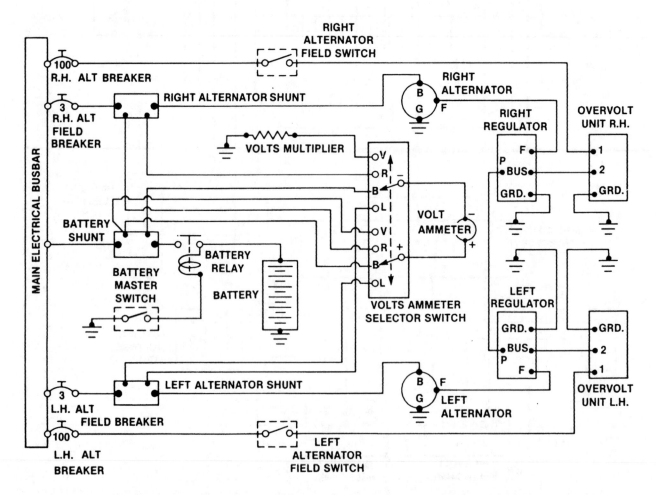

Fig. 2-19 Alternator control balanced regulator system schematic

1. Operation

Turn the battery master switch "on", start engines. Place both alternator field switches to "on" and the alternator will excite and bring the alternator outputs onto the busbar.

The excitation circuit for the alternator is from the field circuit breaker through the alternator field switch into pin 1 of the overvolt unit, out of pin 2 to the bus terminal on the regulator out of the F terminal and through the field windings to ground. Initially, the bus and F terminals on the regulator have a very low resistance between them so that almost maximum field current flows through the regulator to allow rapid build up of alternator voltage.

As the voltage gets up to normal operating level a transistor in the regulator starts to increase the resistance in the field circuit. This transistor senses output voltage and if the voltage rises above the regulated level the transistor increases its resistance and reduces field current. If the output voltage drops below the regulated level the transistor decreases its resistance and increases field current to raise alternator output voltage.

The two P terminals on the voltage regulators are used to connect the two field voltages in the regulators for comparison. If one alternator's output rises compared to the other, its field voltage will rise. This rise in voltage is sensed by the other regulator through the interconnecting wire between the P terminals. This increased field voltage fed to the lower voltage regulator causes a paralleling transistor to adjust the voltage of the lowest voltage regulator back up to the same voltage as the highest regulator.

When both alternators produce the same output voltage, a voltmeter connected between the two F terminals should read 0 volts. Adjustments for the paralleling system are based on adjusting the two voltage regulators until the voltmeter connected as above reads as close to zero as possible.

The overvolt units sense busbar voltages between terminals 1 and ground. If this voltage should exceed specified limits, the normally closed circuit between terminals 1 and 2 will open and remove busbar power from the alternator field.

The volt ammeter selector system is arranged so that if the switch contacts are connected across appropriate circuit points, it will read that particular value.

(a) +L −L reads millivolt drop across left alternator shunt in amps.
(b) +B −B reads millivolt drop across battery shunt in amps.
(c) +R −R reads millivolt drop across right alternator shunt in amps.
(d) +V −V reads voltage between busbar and ground. The volts multiplier resistor increases the range of the millivoltmeter to aircraft voltage.

NOTE: When reading the amps ranges on the voltammeter, the voltammeter is really reading the millivolts drop across the shunt. The meter is marked in amps, but is really a sensitive voltmeter. It is important when setting up the system for parallel operation that the manufacturer's manual procedures are followed.

Review Questions:

1. Explain the purpose of the regulator selector switch.

2. How is the field circuit of a failed alternator isolated from the busbar?

3. If contacts 1, 3, 2 and 6 were selected in the ammeter selector switch, what would the ammeter indicate?

4. If the overvoltage relay senses an overvolt condition, what action takes place?

5. What action could the pilot take to try to remove the problem in question 4?

6. What is the purpose of an R.F.I. filter?

7. What advantage do individually controlled alternators have over the system described in this chapter?

8. What voltage should be present if you connected a voltmeter between the F terminals on the regulator if the system is working correctly?

9. What function does the selection of the fourth position on the volt ammeter selector switch perform?

10. If you connected a voltmeter between the bus terminal of a regulator and ground, what voltage would you expect to read with the alternator field switch (a) closed — (b) open, and the engines stopped?

11. How would you adjust the individual voltage regulators to increase or decrease voltage?

CHAPTER 14

Twin Engine Internal Lighting Systems

Most twin engined aircraft have internal lighting systems which are broken down into various sections which vary according to the aircraft manufacturer.

In general, Cessna divide their internal lighting into the following groups:

Radio Light Circuits
Flight Instrument Light Circuits
Engine Instrument Light Circuits
Flood Light Circuits
Cabin Light Circuits

Piper, however, group their internal light circuits into:

Flood Light Circuits
Avionics Light Circuits
Main, Copilot, Middle, Lower and Autopilot Light Circuits
Forward Baggage Compartment and Passenger Reading Light Circuits

Certain circuits, especially those associated with the illumination of instruments, must have the facility to be varied in light intensity so they are equipped with dimming controls to raise or lower the light levels.

In earlier models of aircraft, the dimming was achieved by the use of rheostats placed in the common feed wire to the various groups of lights.

As the number of instruments and avionics units grew, more lights had to be added with the result that the size of the rheostats had to increase in order to handle the higher loads placed in the circuit by the extra lights.

As most dimming rheostats are mounted on the instrument panels and with their increased size, panel space to accommodate the larger rheostats, increased instruments and avionics became more limited.

In later model aircraft, these problems were overcome by using transistors to drive the lights. The transistors are remotely controlled by very small rheostats located on the instrument panel and the transistors are mounted in locations remote from the instrument panel.

A. Early Model Instrument Panel Light Circuit

1. Operation

With the battery master switch turned to the "on" position, power is available to the main electrical busbar. With the rheostat wiper W on, contact A of the flight instrument light rheostat, no current will flow through the rheostat to the lights.

If the wiper is now positioned to contact B on the rheostat current limited by the resistor D will flow through the lights and they will glow dimly, due to all of the resistance D being in the circuit. As the wiper W is moved towards contact C, less resistance is in the circuit and the lights will increase in brilliance until at point C, the full current will go from the busbar directly to the lights through the wiper and contact D and the lights will be at full brilliance.

The other light circuits shown on Fig. 2-20 operate in a similar manner as described in the above paragraphs, except that they take their power from the 2 amp circuit breaker as shown.

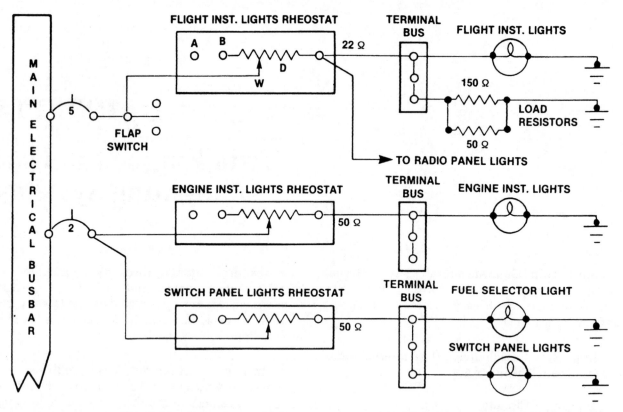

Fig. 2-20   An early model instrument panel light circuit of a Cessna 310

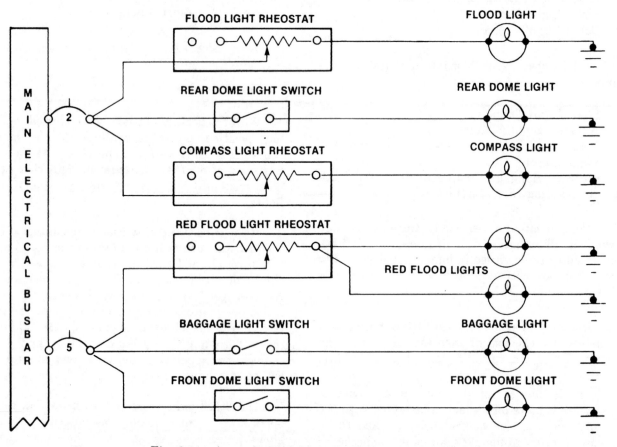

Fig. 2-21   An early model interior light circuit of a Cessna 310

78

NOTE: The load resistors are in the circuit to balance the load if only the flight instrument lights are connected to the terminal bus. If extra radio panel lights are connected to the circuit as shown, the resistor values of the load resistors must be adjusted as required to assure the correct light intensity. The diagram indicates only one light connected to each of the flight instrument and engine instrument terminal busses. In actual practice, as many lights as needed are connected to these points.

B. Early Model Interior Lights Circuit

1. Operation

The circuits controlled by rheostat work the same as previously described in the early model instrument panel light circuit. The front and rear dome lamps and the baggage compartment lights are controlled by their individual switches and burn at full brilliance at all times when selected.

C. Transistor Dimming Control

The more recent aircraft have as mentioned previously, dimming controls which utilize transistors in their operation. To understand how these systems work, it is necessary to know something about transistor operation.

D. The Transistor

A transistor is a semi-conductor device which can be used as a remote controlled switch or a remotely controlled resistor.

A transistor has no moving parts and functions on the principle that when the correct voltage is applied to its control terminal it will allow a greater current to flow between its two main circuit terminals. Transistors are shown in symbolic form by the symbols shown in Fig. 2-22.

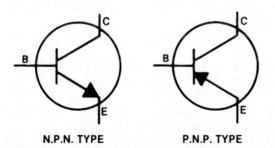

N.P.N. TYPE   **P.N.P. TYPE**

Fig. 2-22 Transistors that are shown in symbolic form by symbols

The terminal identifications are as follows:

B = base or control terminal
C = collector or terminal which accepts main circuit current
E = emitter or terminal which emits main circuit current

The following types signify what polarity of materials are used in the transistor:

N.P.N. = negative collector-positive base-negative emitter
P.N.P. = positive collector-negative base-positive emitter

In general terms, an N.P.N. transistor requires a positive voltage on its base to cause the collector and emitter to conduct; ie., a positive voltage on the base lowers the circuit resistance between the collector and emitter.

The amount of voltage required to start a transistor conducting is very small. A voltage of .3 volts positive applied to the base of an N.P.N. transistor will start the transistor conducting. If this voltage is increased to .5 volts positive, the transistors collector emitter junction will go to minimum resistance; ie., full conduction.

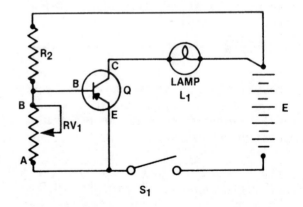

Fig. 2-23 Transistor control circuit

1. Operation

The positive of the battery E is applied to the circuit as follows:

(a) through lamp L_1 to the collector C of the transistor Q
(b) through R_1 to the base B of the transistor Q

79

As the switch S_1 is not closed, no current can flow in the circuit back to the battery E. RV_1 is placed with its wiper to position A and S_1 is placed to the "on" position. Current flows through R_2 and RV_1 back through the switch S_1 to battery negative. The current flow through R_2 and RV_1 creates a voltage drop across RV_1 which is below .3 volts+. This is insufficient to cause the base B of the transistor Q to cause the collector emitter circuit to conduct.

RV_1 wiper is now moved toward point B and the voltage at the base B of the transistor increases to approximately .3 volts positive. The junction of the collector-emitter reduces its resistance and current starts to flow from the battery positive through lamp L_1 through the collector-emitter junction through the switch and back to the battery negative.

As the wiper is moved further toward point B, the voltage on the base increases and causes a further reduction in the resistance of the junction of the collector and emitter. This allows more current flow through the lamp and its brilliance increases.

When the wiper RV_1 reaches point B, the voltage applied to the base is approximately .5 V+ and the junction of the collector-emitter is at its minimum resistance. This allows maximum current to flow through the lamp and it burns at full brilliance.

By providing the correct voltage condition on the base of the transistor, we can vary the collector to emitter circuit resistance which in turn, will vary the current through the lamp to get the desired light intensity.

The current drawn by the base circuit, is extremly small compared to the current which flows from the collector to emitter so one very light small rheostat can control many lights if the lights are connected in the transistor emitter collector circuit.

E. Transistorized Dimming System of a Cessna 310

1. Operation

Each circuit in Fig. 2-24 works on the same principles as follows:

1. Power is applied from the busbar, through the flight instrument circuit breaker to point B on the radio lights dimming control.

2. With the wiper N at point A, the power at the base B of the radio lights control transistor is 0. volts.

3. The power is also routed from point B of the dimming control through pin 1 of the transistor unit plug to the collector C of the transistor. No current flows through the collector C to emitter E and out to the radio panel lights as the base B has no positive voltage applied.

4. As we move the wiper W from point A, this causes a positive voltage to be applied to the base B of the transistor and a small amount of current flows from the collector C through the emitter E through pin 3 of the plug, through the radio lights terminal block through the lights to ground. The lights will illuminate at low intensity.

5. The further we move the wiper A from point A towards point B, the higher the positive voltage on the base B. The collector C to emitter E junction decreases resistance as the base B becomes more positive. The light intensity increases.

6. When the wiper W reaches point B, maximum positive is applied to the base B and the transistor collector emitter junction comes to its lowest resistance. Maximum current flows through the junction out to the lights which will light up to their full brilliance.

7. Selection of the wiper position at the appropriate point between A and B will adjust the lights brilliance to the desired intensity.

8. Each of the transistors and subsequently the groups of lights operate as described early in this chapter.

NOTES: 1) The radio and flight instrument panel lights share a common feed from the flight instrument light breaker. 2) The switch panel and engine instrument lights share a common feed from the engine instrument lights breaker. 3) The transistors are mounted on a heat sink unit and the relevant circuitry is connected by a multi pole plug. This heat sink unit is mounted to the air-

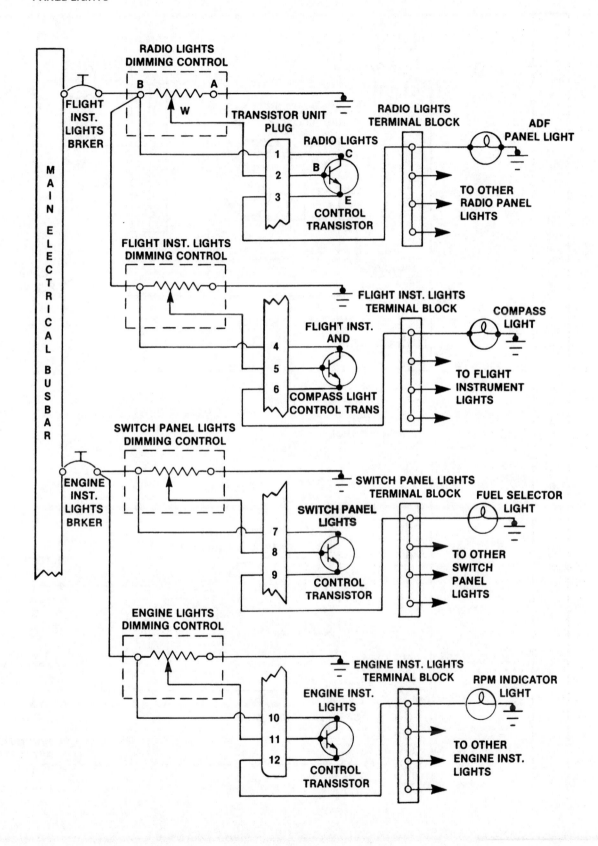

Fig. 2-24   Typical transistorized dimming system

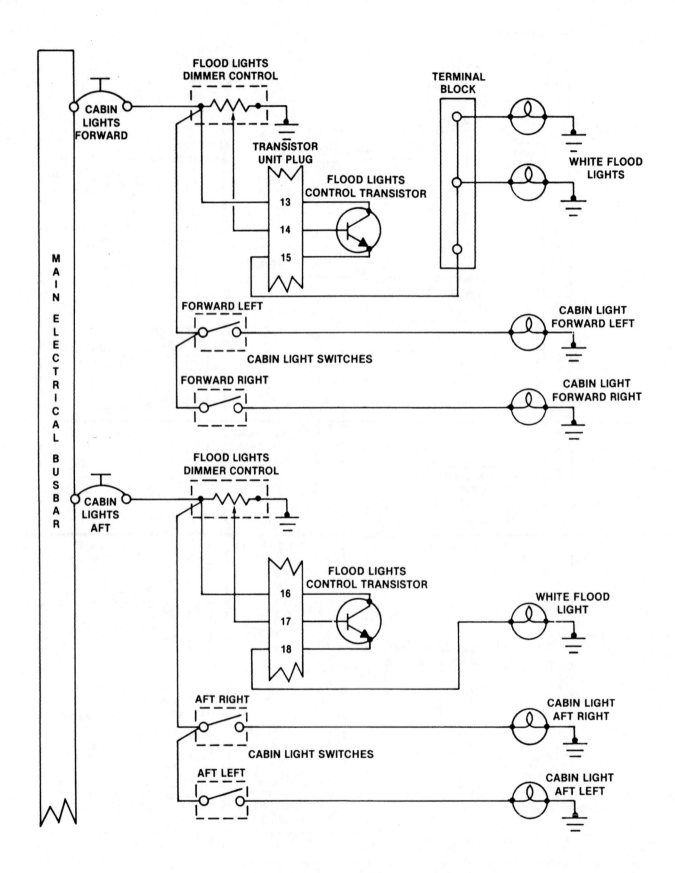

Fig. 2-25 A cabin lighting system for a Cessna 310

frame structure away from the instrument panel. 4) The dimming controls are light and small potentiometers mounted conveniently on the instrument panel.

F. Cabin Lighting System

1. Operation

The dimmer controls work as described in this chapter to operate the flood lights. The cabin lights share a common circuit breaker with the dimmer floodlight controls and are individually selected by the appropriate switch.

Review Questions:

1. How are the instrument lights dimmed in an older type aircraft?

2. In Fig. 2-20 what is the purpose of the two load resistors?

3. Why was it found necessary to use transistor dimming control on later aircraft?

4. The terminal on a transistor which controls the amount of current flow through the transistor is known as:

5. Collector and emitter connections on the transistor carry which current?

6. If we apply a positive to the base of an N.P.N. transistor in a light dimming circuit, what will happen to the emitter collector circuit?

7. What effect would an open ground connection on the flood lights dimmer control have on the brilliance of the floodlights in Fig. 2-25?

CHAPTER 15

Twin Engine External Lighting Systems

The external lighting systems on the aircraft are:

1. Position Lights — Navigation Lights
2. Landing Lights
3. Taxi Lights
4. Rotating Beacons — Anti-collision Lights
5. Strobe Lights — Anti-collision Lights

A. Position or Navigation Lights

The navigation lights are the red, green, and clear lights positioned at the extremities of the aircraft to warn other aircraft of which way the aircraft is turning. The navigation lights are a legal requirement for night flying and the following standards must be met:

(a) angles of divergence of lights
red-port   110°
green-starboard   110°
clear-tail 140°

(b) light intensity
red-port — approximately 20W
green-starboard — approximately 20W
clear-tail — approximately 10W

Navigation lights can be used in two modes which are steady or flashing. The circuit must have the necessary changes to the system wiring to accommodate these options if desired. Some aircraft have only the provisions for steady state navigation lights.

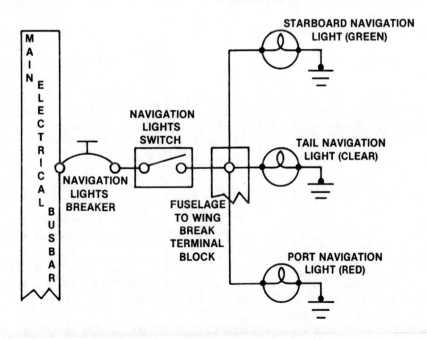

Fig. 2-26 Steady state navigation light circuit

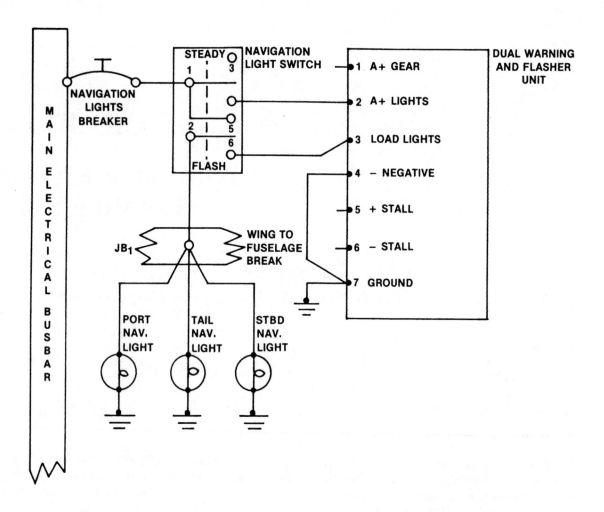

Fig. 2-27 A diagram of a steady and optional flasher navigation light circuit

B. Steady State Navigation Light Circuit

1. Operation

Power from the main busbar is routed through the navigation lights breaker, through the navigation lights switch to the wing break terminal block and then through the lights to ground.

C. Steady and Optional Flasher Navigation Light Circuit

1. Operation

Steady state operation: Place Nav lights switch to steady position. Power flows from busbar through the Nav lights breaker, through the Nav lights switch pins 1, 5, 2, and through JB$_1$ to the Nav lights. The lights burn steady.

Flashing operation: Place Nav lights switch to flash. Power flows through the Nav lights breaker, through the Nav lights swith pins 1 to 4, to the flasher unit pin 1. The flasher unit is energized and provides a switched or flashed output from pin 3 of the flasher unit, through the Nav lights switch pins 6 and 2, through JB$_1$ to the lights. This switched output causes the lights to flash on and off.

D. Landing and Taxi Lights

Twin engine aircraft are produced with two basic types of systems which are:

1. Fixed landing and taxi lights mounted in a transparent cover and usually mounted in the wing.

2. Retractable landing lamp and fixed taxi lamp. The landing lamp would be attached to

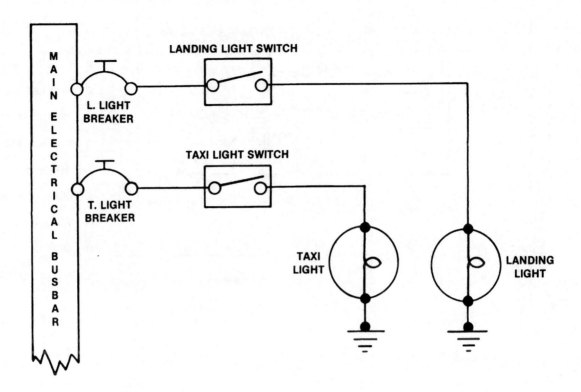

Fig. 2-28 Typical fixed landing and taxi light circuit

the underside of the wing, nacelle or tip tank and the taxi lamp to the nose oleo leg.

1. Operation

The operation of this circuit is very simple. Selection of either switch to "on" feeds power from the busbar through the applicable circuit breaker and switch to the the lamp.

E. Retractable Landing and Taxi Lamp Circuit

The retractable landing lamp has a reversible motor which drives the lamp assembly in and out of the wing as demanded by the selector switch. The drive to the lamp assembly is through reduction gearing.

Most retractable landing lights have a slipping clutch on the gear mechanism which is set to slip if the lamp is extended at a higher speed than called for on the pilot's check list. This prevents damage to the lamp assembly and gearing due to severe slip stream effects.

1. Operation

The taxi light operation is simple, as shown by its circuit. With the switch turned on, power is applied to the taxi light.

The landing lamp circuit is slightly more complex in operation. The lamp must be selected down. The actuator must extend the lamp, the lamp must be switched on and it must stop at the end of its travel down. The reverse must occur when the lamp is selected up.

Components in the lamp assembly are as follows:

A Extend Limit Switch
B Retract Limit Switch
C Extend Field
D Retract Field
E Motor Armature
F Brake Solenoid Coil
G Moving Lamp Contact
H Lamp
I Fixed Lamp Contact

2. Sequence of Operations

(a) Lamp H retracted. Turn selector switch to extend. Power is applied through the closed extend limit switch A, extend field coil C, armature E, brake solenoid coil F to ground. The motor armature runs in an extend direction and the lamp H starts to extend.

87

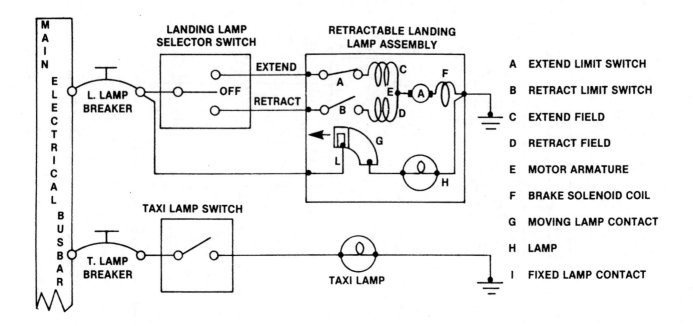

Fig. 2-29 Retractable landing light and taxi lamp circuit for Cessna 310

(b) The moving lamp contact G moves in the direction of the arrow and the fixed contact L moves from the small insulated portion of the moving contact G to the large uninsulated portion of the moving contact. Power is applied from the L lamp breaker through the fixed and moving contacts L and C, through the lamp to ground and the lamp comes on.

(c) As soon as the lamp mechanism moves away from the retracted position, the retract limit switch B closes. This makes both limit switches in the closed condition during transit and ensures that an extend or retract selection can be made should it be necessary.

(d) The lamp runs down to the full extend position and the extend limit switch A opens. This removes power from the extend field C armature E and brake solenoid F. The motor stops and the solenoid brake F contacts the motor shaft to prevent run on.

(e) The lamp is now extended and "on". The extend limit switch A is open and the retract limit switch is closed.

(f) Selector switch is selected to retract. Power is applied through the closed retract limit switch B, through the retract field D, through the armature E and through the brake solenoid F to ground.

(g) The brake solenoid lifts the brake from the armature shaft and the motor turns in a direction opposite to the extend condition. The

lamp H starts to retract and moves the moving lamp contact until the lamp is just about to be fully retracted. At this time, the fixed contact is contacting the insulated portion of the moving contact and the landing lamp H is switched off.

(h) When the lamp is fully retracted, the retract microswitch B opens and switches off the motor.

(i) The lamp is now retracted and "off" the retract limit B switch is open and the extend limit switch is closed, ready for the next extend selection.

F. Rotating Beacon (Red)

The rotating beacon is a high intensity red light which is placed primarily on the fin of the aircraft. Some aircraft have a second beacon installed on the belly of the aircraft in order to ensure that the aircraft is visible from above and below. The beacons come in two types: motor driven reflector and fixed reflector.

In the motor driven type, a small motor rotates the reflector past the bulb and due to the reflector angle causes the light to flash on and off.

In the fixed reflector type, the flash is produced by an electronic flasher unit which flashes the light on and off.

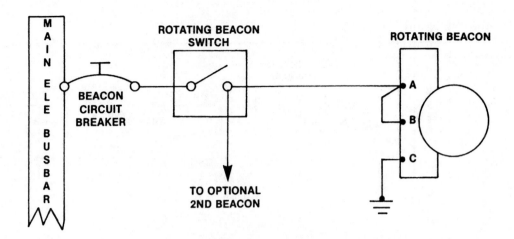

Fig. 2-30   Motor driven reflector beacon unit

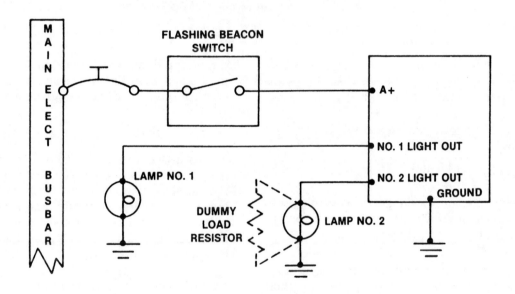

Fig. 2-31   Fixed reflector flashing beacon unit

1. Operation of motor driven reflector beacon unit

The selection of the switch to on applies power to pins A and B. One pin is the lamp supply, the other is for the motor. The motor runs and turns the lens and reflector past the light causing it to flash on and off.

NOTE: Some aircraft have two beacons. They may be controlled by a common switch as shown in Fig. 2-30, or may be a complete duplicate circuit of Fig. 2-30.

2. Operation of a fixed reflector flashing beacon

Selection of the flashing beacon switch to on applies power to the A+ terminal of the beacon flasher unit. This unit provides alternate outputs to No. 1 and No. 2 lights through its electronic flashing action. No. 1 comes on and goes out followed by No. 2 and the action is repeated. The flasher continues to supply the lights alternately as long as the switch is on.

NOTE: The resistor shown dotted and identified as the dummy load resistor is substituted

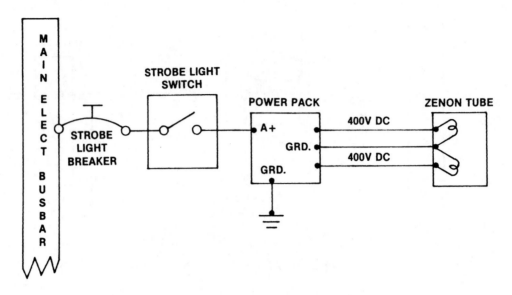

Fig. 2-32 Typical strobe light schematic

for the No. 2 lamp when only No. 1 lamp is install-ed on the aircraft. This ensures that the power produced at the No. 2 light output is dissipated across the dummy load resistor to minimize puls-ing on the ammeter due to an unbalanced load condition.

G. Strobe Lights

Another type of flashing beacon coming into more frequent use on many general aviation air-craft, is the high intensity strobe light.

This light emits a brilliant white light of ap-proximately 1 million candle power in short flashes. The high intensity white light is caused by current discharge through a zenon tube filled with a conducting gas.

To cause current to flow through the gas and emit light, a very high DC voltage of approxi-mately 400 volts DC is required. This is supplied by a strobe light power pack situated close to the zenon tube.

The power pack accepts 14 or 28 volts from the aircraft system and converts it to high voltage DC which it stores in a capacitor. When the capacitor becomes fully charged to 400 volts, the zenon tube strikes or fires the stored charge in the capacitor, rushes through the gas in the zenon tube and emits a high intensity white light. This process is repeated at frequent intervals to cause repetitive flashing.

Strobe lights may be fitted on the fin or on the fin and wing tips of the aircraft so there may be more than one light on a given aircraft.

1. Operation

Turn the strobe light to on. Aircraft voltage is applied to the power pack. Voltage output of the power pack builds up to 400 volts DC on each line and discharges through the zenon tube to ground. This causes a brilliant white flash from the tube. This operation is repeated approximate-ly at 1 second intervals.

NOTE: The voltage at the strobe light tubes is 400 volts DC and should be treated as dangerous if the lights are selected on. Do not handle the tubes with bare hands while changing tubes as oil from the hands will cause local heat spots and tubes may fail prematurely.

Review Questions:

1. Name four common external lighting sys-tems.

2. The position or navigation lights require 360° coverage in the lateral plane of the aircraft. How is this achieved?

3. How is the flashing state achieved on the pos-ition lights in Fig. 2-27?

4. What safeguard is incorporated to prevent damage to the mechanism of a retractable landing lamp?

5. What switches the lamp off when it is retracted?

6. What drives the lamp down and up?

7. What prevents over-running of the mechanism of a retractable landing lamp?

8. What is the purpose of the dummy load resistor in Fig. 2-31?

9. What voltage is required to work a strobe light?

CHAPTER 16

Twin Engine Landing Gear Control and Indication Systems

Many light twin engined aircraft have retractable systems. Each of these systems require two basic subsystems which are:

1. Landing gear selection and operation.
2. Landing gear position indication.

There are three common formats for landing gear control and indication systems. They are:

Format #1 — Hydraulic control and operation of the landing gear and electrical indication of its position.

Format #2 — Electrical control and operation of the landing gear and electrical indication of its position.

Format #3 — Electro-hydraulic control and operation of the landing gear and electrical indication of its position.

A. Types of Gear Position Indicating Systems

Most electrical gear position indicating systems work on a basic principle. When the gear reaches a pre-determined position, a microswitch is operated by a cam or lever and the microswitch then completes a circuit to give a light indication to verify the gear position.

The majority of gear position indicating circuits have two microswitches for each leg of the gear system. These microswitches are almost universally referred to as the downlock and uplock microswitches.

The number of lights used to indicate the gear position varies depending on the manufacturers preference, but virtually all systems use the previously mentioned two microswitches per leg of the landing gear.

The factor which determines how many lights are used is the method of connecting the microswitches to the lights. If the switches are connected in series to the lights, only two lights are used. Green for gear locked down, and red or amber for gear locked up. If the switches are connected from the busbar in parallel to the lights, six lights may be used—three green and three red. Some systems use three green and one red light so the green lights are connected between each downlock microswitch and ground and the red light is connected in series with the three uplock microswitches.

Landing gear indication and warning systems usually include an audible warning system such as a horn or klaxon which is activated by a microswitch on the throttle linkage. When any or all of the gear legs are not locked down and the throttle is retarded to less that 1/3 full power the horn sounds due to the operation of the throttle microswitch.

1. Operation of Fig. 2-33 — gear locked down

When the gear is driven down and as the locks come on, the cams on the gear legs operate the respective leg downlock microswitches. The switch levers in each downlock microswitch are moved so that contacts C and N.O. are connected.

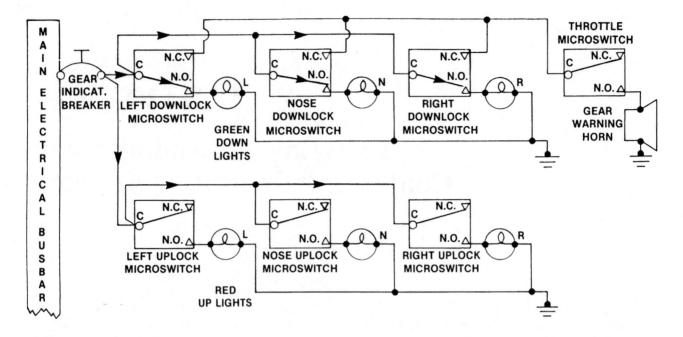

Fig. 2-33 *Landing gear position indicating circuit. Three green, three red lights: gear locked down, throttle open.*

This allows power as shown by the arrow head symbols to be applied from the gear indication breaker to the C contacts on each downlock microswitch through each switch lever and N.O. contact to each green down light. Each green down light will illuminate indicating that each gear leg is locked down.

It will be noted that although power is also applied to the C contacts of the uplock microswitches (shown in double arrows), it is not applied to the red up lights as the microswitch levers are bridging from C to N.C. contacts in each switch which is, of course, open circuited. The only time the microswitch levers are changed to bridge the C to N.O. contacts are when the uplocks are locked.

As the downlock microswitches are operated, the circuit between C and N.C. on each switch is open so regardless of throttle microswitch position no power is available to sound the warning horn.

2. *Operation of Fig. 2-34 — gear locked up, throttle retarded*

As the uplocks come on, the cams on the gear legs operate the respective uplock microswitches.

The microswitch levers in each uplock switch are moved so that contacts C and N.O. are connected.

The downlock microswitch levers are not operated so the downlock switch levers are connected from C to N.C. The throttle microswitch lever is bridging contacts C to N.O.

Power is applied from the gear indication breaker along two paths as shown by the arrow heads. The first path goes through the downlock microswitches and the throttle microswitch to the gear warning horn which sounds, indicating that the gear is not locked down. The second path goes through the uplock switches contacts C to N.O. and the red up lights illuminate. If the throttle is advanced above the 1/3 open position as in normal flight, the throttle microswitch lever will go to the C to N.C. position and the horn will not sound.

3. *Operation of Fig. 2-35 — gear in transit*

When the gear is neither locked up or down, but is traveling in either direction, none of the leg microswitches are operated. Each microswitch is, therefore, in the C to N.C. contact position. This breaks the circuit to both the red and green lights.

94

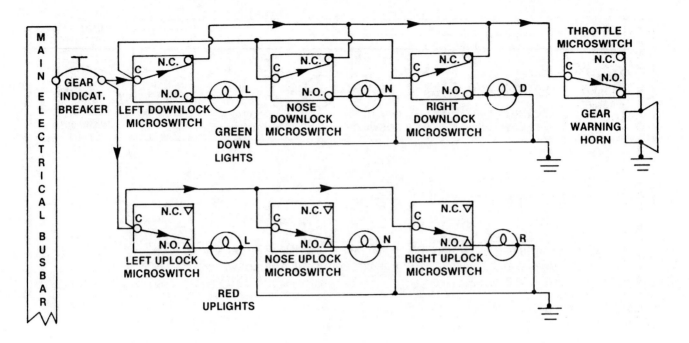

Fig. 2-34 Landing gear position indicating circuit. Three green, three red — gear locked up, throttle retarded to less than 1/3 open.

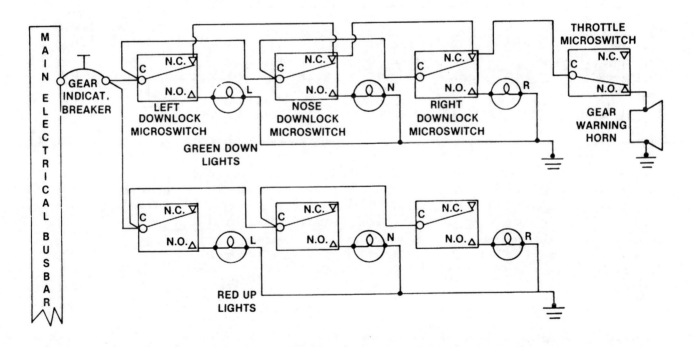

Fig. 2-35 Landing gear position indicating circuit. Three green, three red — gear in transit, throttle retarded to less than 1/3 full power.

Power is, however, allowed through the C to N.C. contacts on the downlock microswitches to the throttle microswitch. In the position shown with contacts C to N.O. connected, the horn will sound.

Advancing the throttle to more than 1/3 open will operate the throttle microswitch and changes its contacts so that C to N.C. will be connected. This opens the circuit to the horn and the horn stops.

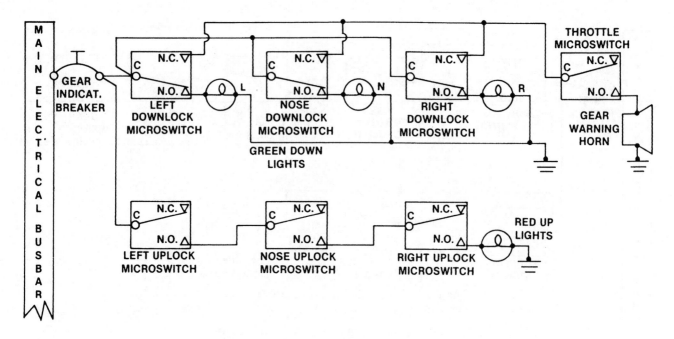

Fig. 2-36   *Typical landing gear position indicating circuit. Three green, one red.*

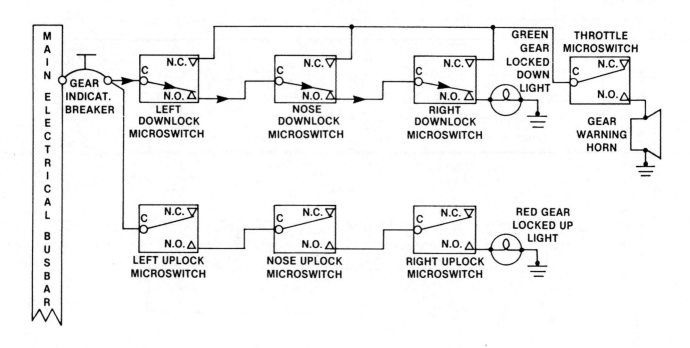

Fig. 2-37   *Gear position indicating system. One green, one red — gear locked down.*

4.   *Operation of Fig. 2-36 — gear locked down*

The operation of this circuit in the gear locked down position is identical with that in Fig.
2-34.

5.   *Fig. 2-36 — gear locked up*

In this circuit, it should be noted that the only way to get the red light on is to operate all uplock microswitches to the lower C to N.O. contact

position. If one switch does not operate, the red light will not come on. The horn operates in the same manner as described in Fig. 2-34.

6. Fig. 2-36 — gear in transit

In this position, none of the microswitches are locked in any position, therefore, no lights are on. All microswitches are connected to C to N.C. so that the down microswitches supply the horn circuit through the throttle microswitch when the throttle is retarded.

B. Gear Position Indicating System

1. Operation of Fig. 2-37 — gear locked down

Fig. 2-37 shows the path of power applied to the green down light with all downlock microswitches operated by the downlock cams. The arrows indicate the circuit path. It should be noted that with the microswitches in the positions shown that no power is available to either the throttle microswitch and horn or to the red gear up light.

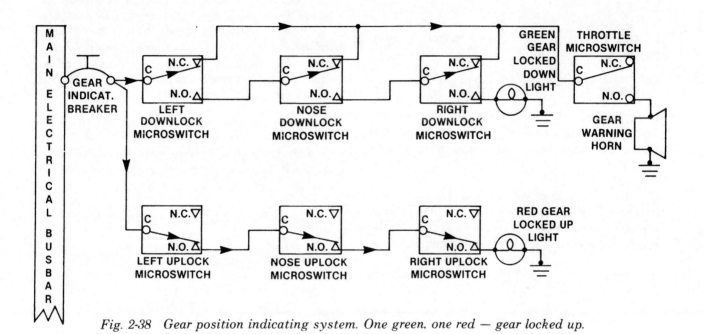

Fig. 2-38 Gear position indicating system. One green, one red — gear locked up.

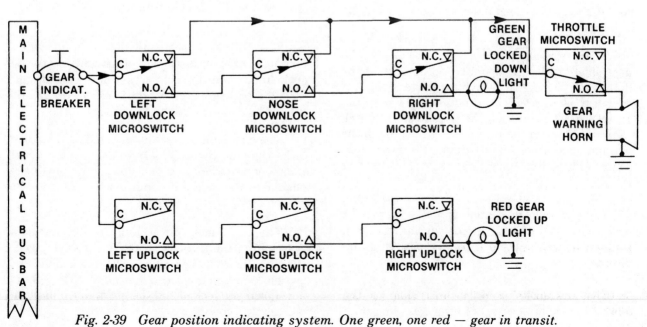

Fig. 2-39 Gear position indicating system. One green, one red — gear in transit.

2. Operation of Fig. 2-38 — gear locked up

Fig. 2-38 shows the path of power applied to the red down light and the throttle microswitch shown by the arrows.

If the throttle is retarded to less than 1/3 full throttle, the throttle microswitch changes over and applies power to the warning horn that the gear is not locked down.

NOTE: If any downlock microswitch is in the position as shown, power is available to the throttle microswitch.

3. Operation of Fig. 2-39 — gear in transit, throttle retarded

Fig. 2-39 shows the path of power to the gear warning horn which operates and indicates gear not locked down. The path is shown by arrows. The circuits to the lights are not connected through the microswitch contacts and both lights are out.

It should be once again noted that any downlock switch which is unlocked will apply power to the throttle microswitch so that selection of the throttle to less than 1/3 open will sound the horn.

4. Operation of Fig. 2-40 — selected up from down

In Fig. 2-40 where the switches are shown in solid lines, is the gear locked down position. The dotted switch positions indicate a gear locked up condition.

When the aircraft becomes airborned, the gear safety switch is operated and moves to the dotted line condition. Moving the gear selector switch to the up selection routes power from the 15 amp breaker through the gear selector switch to terminal A_2 on the gear relay then through the up limit switch (solid line) through the gear safety switch (dotted line) to energize the coil on the gear relay.

The gear relay closes and applies power to the up terminal of the gear motor causing it to run through its gearing and drive the gear in an up direction.

The downlocks unlock which changes the down limit switch contacts to the dotted position and the nose, left and right gear down indicator contacts to the dotted positions. The motor circuit is now prepared for the next down selection and the green down light is extinguished.

The lower portion of the nose gear down indicator switch also routes power from the 3 amp breaker through its dotted position to the throttle switches for use in the warning horn circuit.

As the gear reaches its up limit, the up limit switch is operated by the uplock cam and moves both portions of the up limit switch to the dotted line position. The red up light is connected to power from the 3 amp breaker and illuminates. The bottom portion of the up limit switch breaks the circuit through the gear safety switch to the gear relay coil and de-energizes the gear relay. The gear motor stops and the gear is locked up with a red up light indication.

If the throttle switch is closed by retarding the throttle, power is applied from the 3 amp breaker through the nose gear down indicator switch dotted line condition, throttle switch and flasher to sound the warning horn.

5. Operation of Fig. 2-40 — selected down from up

The switches are in the positions shown by the dotted lines with the gear up and locked.

The gear selector switch is moved to down. Power is supplied through the gear selector switch (dotted line) to the down terminal of the gear motor. The motor turns and starts to drive the gear down.

The uplocks release and the up limit switch operates changing both sets of its contacts to the solid line condition. This action puts out the red up light and prepares the gear relay coil circuit for the next up selection.

The gear continues to extend and as the down locks come on the nose, left and right gear down indicator switches and the down limit switches change to the solid line condition. The down limit switch stops the gear motor.

The nose gear down limit switch isolates power from the throttle switch and routes power through the left then right gear down indicator switch to put on the green down light. The gear is now down and locked and the green down light indicates this condition.

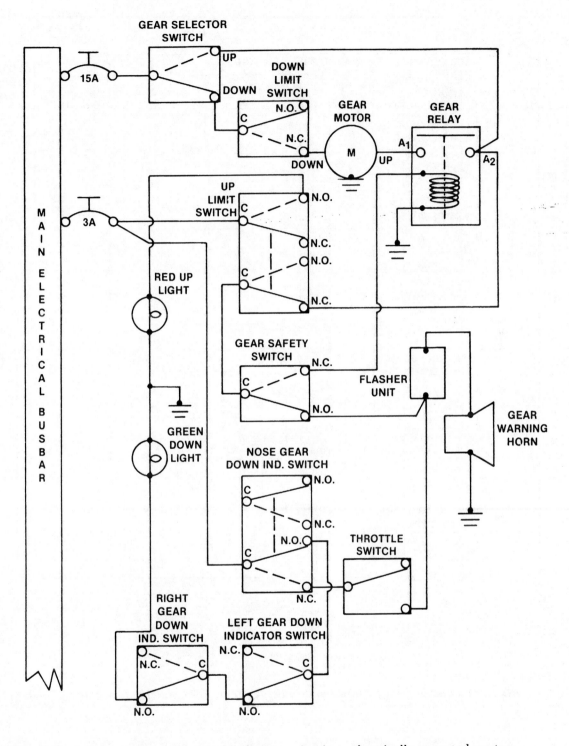

Fig. 2-40 Landing gear control and indication circuit — electrically operated system

As the aircraft lands, the gear safety switch isolates the gear relay coil circuit from power should an inadvertant "up" selection be made with the aircraft on the ground.

6. *Operation of Fig. 2-41 — principles of operation*

This system used electrical control circuitry to run a reversible hydraulic pump motor. The

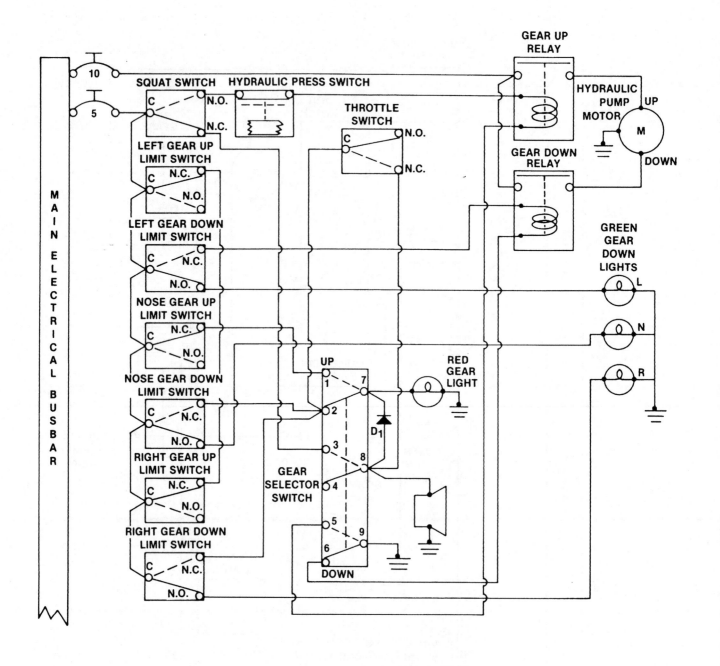

Fig. 2-41 Landing gear control and indication circuit — electrically operated system

motor is controlled by an up and down relay which in turn, are controlled by the hydraulic pressure switch and the left gear down limit switch.

The light indications on this system are slightly different to previous systems described in this section. Three greens indicates the three gear legs are down. One red light indicates that the gear is in transit. No lights indicates that the gear is up. When the throttle is closed with the gear up, the horn sounds and the red light will come on.

The circuit is shown with switches in solid lines in the down and locked condition. The up and locked condition is shown by the dotted line switch positions.

7. Fig. 2-41 — selected up from down

The gear locked down switches in solid line position. Aircraft is airborne. Squat switch changes from N.C. to N.O. position, gear selector switch selected to up. This changes the gear selector switches to the dotted line position. This completes the circuits through the gear selector switch from contacts 1 to 7, 3 to 8, and 5 to 9.

Contacts 5 to 9 complete the gear up relay coil circuit and energize the gear up relay through the squat switch, the normally closed contacts of the hydraulic pressure switch, the up relay coil and contacts 5 to 8 of the selectro switch. The gear up relay contacts close and apply power from the 10 amp breaker to the up terminal of the hydraulic pump motor. Contacts 1 and 7 of the selector switch put the red light on.

The hydraulic pump applies hydraulic power to the gear leg actuator and the gear starts to move up. The gear down limit switches are released and change over to the dotted line positions and the green down lights go out. At this time, the left gear down limit switch prepares the down relay coil circuit for use by connecting power to the coil relay circuit through the C to N.C. dotted connection on the switch. The other two down limit switches connect power to contact 2 on the gear selector switch for use in the horn warning circuit.

The gear legs continue to move upwards until the uplocks come on. At this point, the gear up limit switches change over to the dotted line position. This removes power from contact 1 of the gear selector switch and the red light goes out. Pressure continues to build up in the system as the hydraulic pump motor is still running, but when the pressure reaches a predetermined level, the hydraulic pressure switch opens to the dotted line position and de-energizes the gear up relay switching off the hydraulic pump motor. The gear is now up and locked and no lights are on.

If the throttle is retarded below 1/3 full open, the throttle switches to the dotted line position. This applies power to the horn from contact 2 of

the gear selector switch through the throttle microswitch to contact 8 of the gear selector switch and to the horn. Power also is applied to the red light through the diode D_1 to contact 7 and then to the light. The horn sounds and the red light comes on.

8. Fig. 2-41 — selected down from up

The gear locked up switches in dotted line condition except throttle switch. No lights on. Gear selector switch selected down. Switches move to solid line position.

Contacts 6 and 9 of the gear selector switch complete the ground for the gear down relay circuit and the gear down relay is energized by power from the left gear down limit switch through the coil and pins 6 and 9 of the selector switch to ground. The gear down relay closes and power is applied from the 10 amp breaker through the gear down relay contacts to the down terminal of the hydraulic pump motor. Contacts 2 and 7 of the gear selector switch put the red light on.

The motor drives the pump, the gear starts to extend and the up limit switches changes over to their solid line position. This puts power onto contact 1 of the gear selector switch ready for the next up selection. At this point, the hydraulic pressure switch contacts close to prepare the circuit for the next up selection.

The gear continues to travel down and the down locks come on. The gear down limit switches change over to their solid line position and switch on the green down lights and switch off the red light. At the same time, the left gear down limit switch breaks the gear down relay circuit and switches off the hydraulic pump motor. The gear is now down and locked and the green lights are on.

During transit of the gear the throttle microswitch still controls the operation of the horn as power is applied to contact 2 of the gear selector switch until the nose and right gear down limit switches reach the locked down condition. When the aircraft lands, the squat switch operates and isolates the up relay circuit to prevent an up selection on the ground.

NOTE: When the power is applied to the red light in selected down and travelling condition, the diode D_1 prevents this power being applied to the horn.

Review Questions:

1. In a landing gear indication system with 3 green and 3 red lights, how are the lights connected to the uplock and downlock microswitches?

2. What warns the pilot that his landing gear is not locked down?

3. How are the downlock microswitches changed from an N.C. to an N.O. condition?

4. In the circuit in Fig. 2-37, what conditions must be met to put on the red up indication light?

5. In an electrical system such as Fig. 2-40, what drives the gear up and down?

6. In the same circuit, if the gear fails to extend when selected down, what would be the most likely electrical fault causing this contition?

7. What is the purpose of the diode D_1 in the circuit on Fig. 2-41?

CHAPTER 17

Twin Engine Flap Control System

Electrically operated flap control systems are usually comprised of a reversible DC motor which drives the flap mechanism up and down through suitably designed reduction gearing. The flap mechanism has stops which operate limit switches to stop the flap travel at the extreme ends of the range. The operation of the flap motor is controlled by a flap selector switch which has three positions up, "off" and down. Quite often, the down position on the selector switch is spring-loaded to the centre off position while the up position is not. This is to allow the pilot to inch the flaps down to the desired position but to select full up when he requires it in the event of an aborted landing and go around. Most flap systems also incorporate a flap position indicator system to show the pilot the precise amount of flap selected.

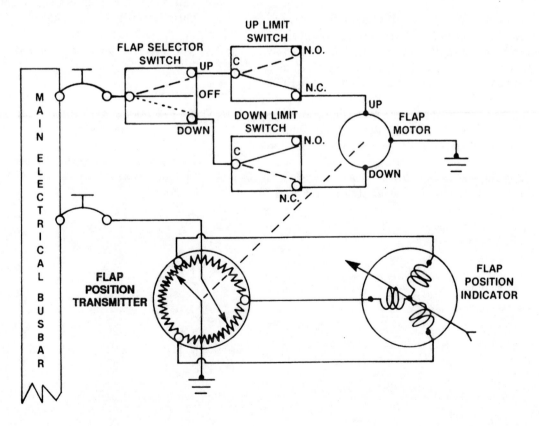

Fig. 2-42   Typical flap control and indication system

103

A. Operation

1. Flap control, flaps selected up

Flaps are in down position as shown in circuit. Selection of flap selector switch to up allows power to be applied to the motor up terminal through the closed contacts of the up limit switch.

The motor runs in an up direction and the flaps start moving up. As soon as the flaps leave the down position the down limit switch changes over its contacts preparing the circuit for a down selection as shown dotted. The flaps continue to move and when they reach the fully up position, the up limit switch changes its contacts and switches off the motor as shown dotted. The flaps are now in a fully up condition.

NOTE: At any point between up and down, both limit switches are closed. This allows the pilot to stop and start the flaps in either direction from any intermediate point.

2. Flap control, flaps selected down

The flaps are up. The limit switches are in the dotted line positions. Selection of the flap selector switch to down applies power through the down limit switch to the down terminal of the motor. The motor runs in the opposite direction to the flap control, flaps selected up, and the flaps move down. As soon as the flaps leave the up position, the up limit switch changes over to the solid line position in preparation for the next up selection.

The flaps continue down until they reach fully down and the down limit switch is operated to the solid line position. This opens up the line to the motor and the motor stops. Placing the flap selector switch to off at any time will stop the flaps at any desired position.

3. Flap position system

As the flap motor is operated there is a mechanical linkage to the flap operating shaft which turns the arrowed wipers in the flap position transmitter. This varies the ratio of electrical power applied to the three coils in the position indicator and lines up the needle in a new position.

The variation in ratio of power applied to the three coils is calibrated on a scale in degrees of flap movement. By monitoring the amount of movement on the indicator the pilot can select precisely the amount of flap required for a particular flight requirement.

Review Questions:

1. What type of switch is used in the flap motor control circuit in Fig. 2-42?

2. How is the flap motor stopped at the end of its travel?

3. If the up limit switch was out of adjustment and did not operate in the flaps up position, what would happen?

4. If the ground on the flap motor became more open circuited, what effect would this have on the operation?

CHAPTER 18

Twin Engine Fuel
Booster Pump Circuits

Many aircraft have electrically drive fuel booster pumps installed. The purpose of these pumps is to provide fuel pressure in an emergency should the engine drive pump fail and to provide fuel priming during start.

In earlier aircraft, the fuel booster pump was simply an electrically driven pump controlled by a simple "on-off" switch. As aircraft became more sophisticated, there became a need for various speeds to be produced by fuel booster pumps to deliver various pressures under different circumstances.

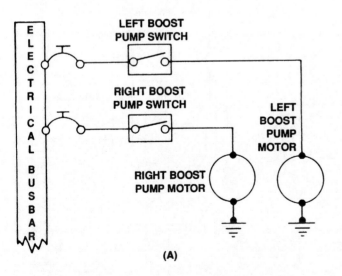

Fig. 2-43 Simple booster pump circuit

A. Operation of a Simple Booster Pump Circuit

The operation of this system is very simple. Selection of the appropriate pump switch will operate that pump at one speed only.

1. Operation of a booster pump 2-stage throttle controlled

Selection of selector switch to low position ensures that power is applied through the speed control resistor to the motor making it operate at low speed.

When the engine is running, the selector switch is placed to "hi". This routes power through the throttle microswitch. If the throttle is set below 1/3 full open, the power must go through the C to N.C. contacts and the speed control resistor and the pump motor runs at low speed.

If the throttle is advanced above 1/3 full power, the C to N.O. contacts of the microswitch are connected. This applies full power to the bottom of the speed control resistor and the pump motor runs at full speed.

B. Functional Description of the System in Fig. 2-45

Priming the engines. Selection of the prime switch to the appropriate engine connects power to the appropriate pump through the particular dropping resistor for that pump. The pump runs at low speed for priming.

When the engines are started the oil pressures build up and both oil pressure switches close and apply power the the two auxiliary pump switches. Selection of the auxiliary pump switches to low passes power to the pump relays and up through the dropping resistors to the pumps, running the pumps at low speed.

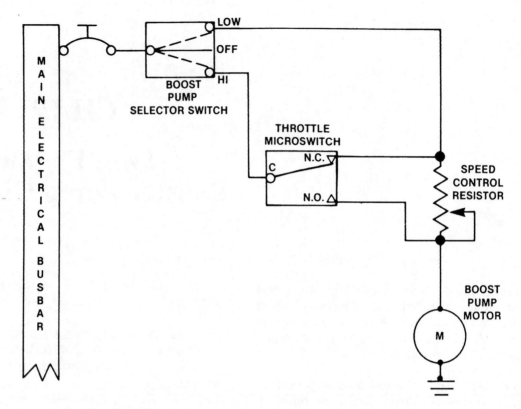

Fig. 2-44   Booster pumps 2-stage throttle controlled

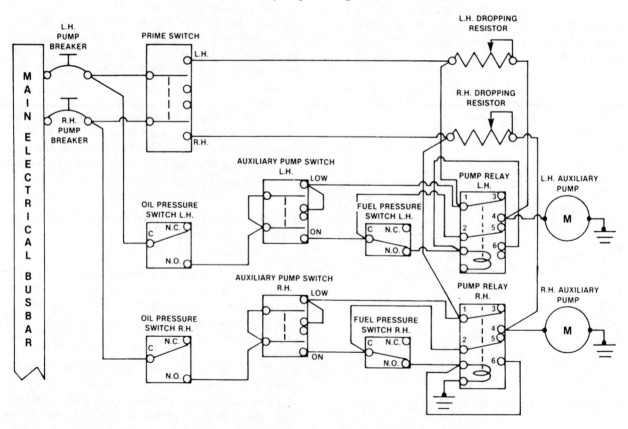

Fig. 2-45   Auxiliary fuel pumps for a Cessna 310

Selection of the auxiliary pump switches to "on" applies power to the fuel presure switches and contact 2 of the pump relays. Power is also applied to contact 1 of the pump relays from the low position of auxiliary pump switches. The pumps continue to run at low speed.

As long as there is fuel pressure in the system the fuel pressure switches will remain open. If the fuel pressure drops to a level which may cause problems in the transfer of fuel to the engines, the pressure switches will close from C to N.C. and energize the pump relays. This applies power across pins 1 and 4 of the relays straight to the motors and they run at full speed. At the same time, power is applied from the C terminal of the fuel pressure switch through contacts 2 and 6 of the pump relays to the relay coil which latches the pump relay on. This ensures that the pump motors continue to run at full speed.

If the fuel pressure problem comes back within normal limits, the pilot may re-position his auxiliary pump switches to off then back to on and the pumps will again run at reduced speed.

The normal position of the auxiliary pump switches for cruise would be "on" and the fuel pressure switches detect loss of pressure. By energizing the pump relay, they correct the situation automatically by increasing pump delivery pressure.

Review Questions:

1. What is the purpose of a fuel booster pump?

2. When will the 2-stage throttle controlled boost pump be running at high speed?

3. What is the purpose of the fuel pressure switches in the auxiliary fuel pump circuit Fig. 2-45?

CHAPTER 19
Troubleshooting

The aircraft maintenance technician's life is constantly made difficult by having to troubleshoot systems with which his training and experience may be hazy.

In electrical systems it is important that before troubleshooting begins, the engineer has a good basic understanding of the system. It is also important that the trouble report given by the flight crew is understood.

Before commencing a troubleshooting assignment the engineer should go through a logical check list of what he needs to successfully achieve a positive result.

A. Steps in Troubleshooting Techniques

1. Confirm the fault as stated by the flight crew.

2. Research the system at fault by reference to manuals and wiring diagrams.

3. Select the test equipment that is required and check that it is working correctly.

4. Proceed to troubleshoot the problem using a step by step approach and constant reference to the wiring diagram.

B. General Troubleshooting of Distribution Circuits

Distribution circuits are comprised of the following basic parts:

1. Busbar connections
2. Fuse or circuit breaker projection
3. Control switching

4. Interconnecting wires
5. A load dissipating component or components

In order to troubleshoot any of these distribution circuits it is necessary to follow a sensible logical procedure to determine the following facts:

1. Is power available at the electrical busbar
2. Does the power get past the protective circuitry
3. Once past the protective device does the power pass through the switching control
4. Having finally passed the switching control does the power arrive at the component.

If the answer to all the above facts is yes the component or its ground return path is at fault.

If power is not available at any of the above mentioned points then further checks are required using appropriate test equipment. If the fault is failure to function completely then the most likely equipment you need is a V.O.M. set on the appropriate voltage scale.

C. Troubleshooting Distribution Circuits With A V.O.M.

If we have a distribution circuit which is defective and is not working at all, then one type of procedure to follow may be:

1. Confirm the fault as reported
2. Refer to the circuit diagram and review how the circuit works
3. Select a V.O.M. to the appropriate range for circuit voltage
4. Proceed next to troubleshooting sequence paragraph below.

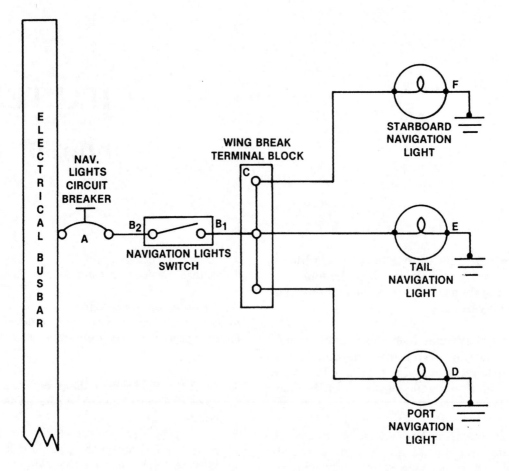

Fig. 2-46 Circuit diagram of a typical distribution circuit

1. Troubleshooting sequence — reported fault all nav lights out — 14 volt system

(a) Confirm fault by selecting nav lights switch to on with master switch on and observe that lights do not come on.

(b) Check that nav lights circuit breaker is pushed in.

(c) Check power to busbar. Select other circuits connected to the busbar. If they function power is on the busbar.

(d) Connect a V.O.M. selected to the 50 volt DC range between point A on the circuit breaker and ground. With the battery master swith "on" the V.O.M. should read 12 volts. If not change the breaker. If Yes proceed to (e).

(e) Connect the V.O.M. between point B_1 on the nav lights switch and ground. With battery master and nav lights selected on the V.O.M. should read 12 volts. If not, connect V.O.M. between point B_2 on the nav light switch and ground and V.O.M. should read 12 volts. If Yes, change nav light switch. If not, change wire between circuit breaker and nav lights switch.

(f) If the original reading at B_1 of the nav lights switch was 12 volts proceed to (g).

(g) Remove nav light lens covers and check bulbs. If all bulbs are defective replace and recheck function of the circuit. If circuit functions generator or alternator output voltage should be checked before flying the aircraft. If all bulbs are serviceable proceed to (h).

(h) Connect V.O.M. between point C on the wing break terminal block and ground with the battery master and nav light switches on the V.O.M. should read 12 volts. If No, change the wire between point B_1 on the nav lights switch and point C on the wing break terminal block. If Yes, check the three wires to the lights and the grounds on the light sockets for serviceability.

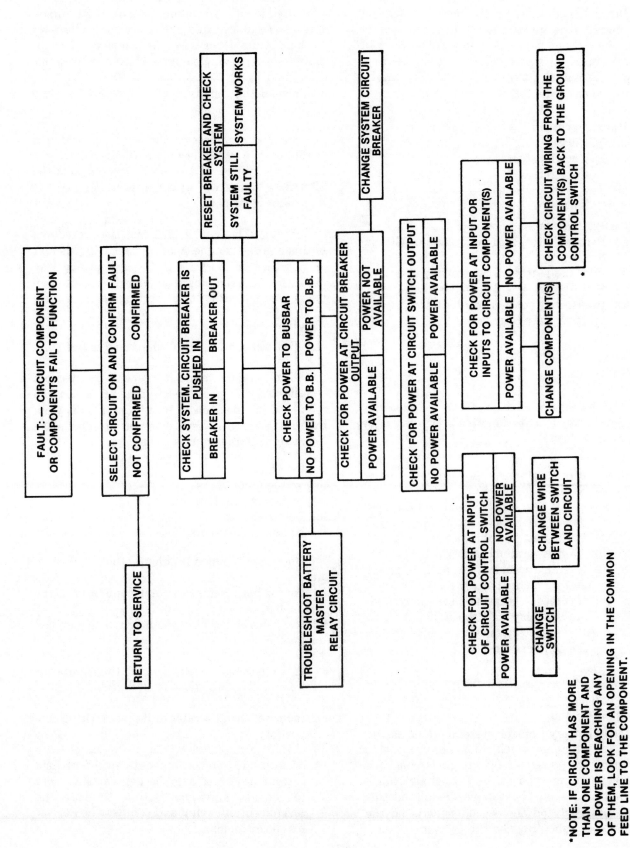

Fig. 2-47 Sample troubleshooting flow chart of distribution circuits

D. Fault Diagnosis Power Generation and Battery Control System

When troubleshooting the power system on the aircraft it is important to break the system down into component parts for the purpose of understanding the problem. The total electrical supply system on a twin engined aircraft can be broken down to the following functions:

1. Battery system:
   (a) Battery
   (b) Battery master switch
   (c) Battery relay
   (d) The system ammeter

2. Generation systems:
   (a) Port and starboard generators or alternators
   (b) Port and starboard controlling devices
   (c) Port and starboard control switches
   (d) Interfacing connections between the two systems were installed

E. Troubleshooting on the Battery Control System

The battery control system is relatively straight forward in its function and reasonably easy to troubleshoot.

1. Start with the obvious and progress to the less obvious.
   (a) A report of no battery power to the bus-bar should raise the following questions:
      i) Is the battery terminals serviceable?
      ii) Are the terminals corroded?
      iii) If both the above are O.K. is power getting to the battery relay?
      iv) If Yes, is the master switch grounding the relay coil correctly?
      v) If the battery relay is clicking in, are the contacts defective?

1. Procedure for troubleshooting a battery control circuit

Most battery control system problems display two symptoms which are no power reaching the busbar or power always on the busbar. The most common method of the first troubleshooting problem is on the sample troubleshooting chart, Fig. 2-48 and the second fault is on the sample troubleshooting chart, Fig. 2-49.

F. Troubleshooting The Alternator

In twin engine electrical power systems we have two identical systems feeding the common busbar. Troubleshooting of individual alternator system faults is exactly the same as that for troubleshooting single engine alternator systems. Fig. 2-50 is a typical sample flow chart for troubleshooting individual alternator system faults.

Some alternator systems work on a shared regulator. If a regulator fails both alternators will fail. If an individual alternator system fails the alternator or the wiring which is unique to that alternator is suspect.

It should be emphasized that the procedures outlined in the troubleshooting charts in this text are samples and are not to be used in any way other than as examples. The aircraft service manual is the final authority on setting up procedures at all times.

G. Troubleshooting DC Generator Systems

Twin engine aircraft generator systems whether they be electromechanical or carbon pile regulated are the same in basic concept. They are both self excited DC systems and both have the following essential components:

(a) generators
(b) voltage regulators
(c) reverse current relays
(d) field switches
(e) equalizing systems

The most common problems which will occur are:
(a) one or both generators not coming onto the busbar
(b) one generator taking too much of the applied load

If we look first at problem (a), the significant failures which could cause such a problem are:

(a) failure of the generator or its main wiring connections
(b) a voltage regulator which in its components or associated wiring prevents sufficient field current from reaching the generator in order to produce sufficient output to close the reverse current relay and connect the generator to the busbar.

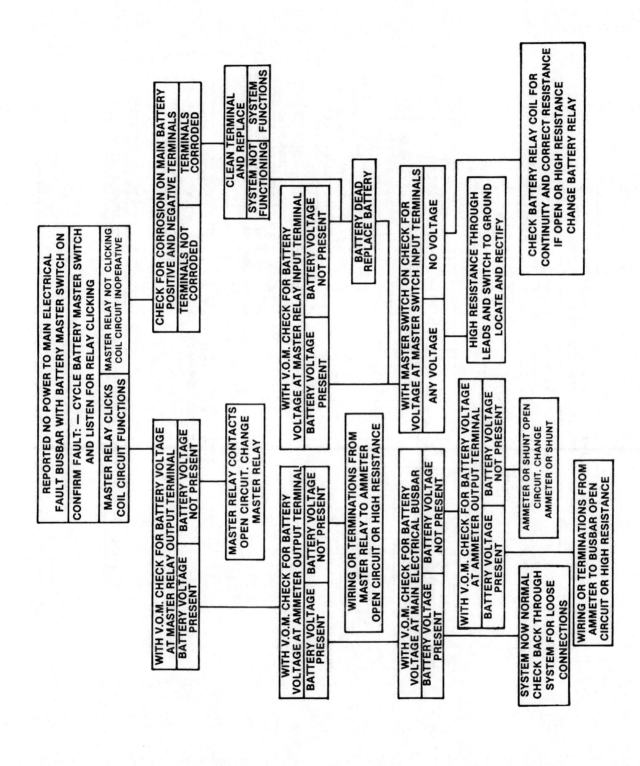

Fig. 2-48 Sample troubleshooting flow chart for a battery control system

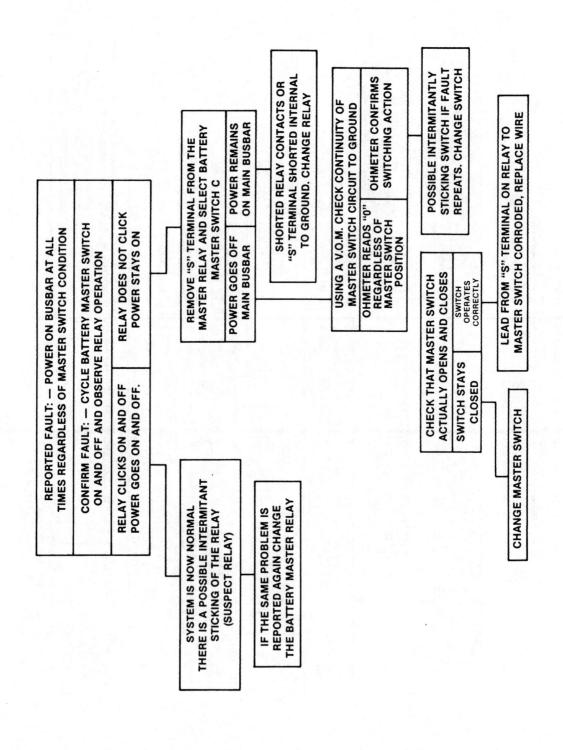

Fig. 2-49 Sample troubleshooting flow chart for a battery control system

114

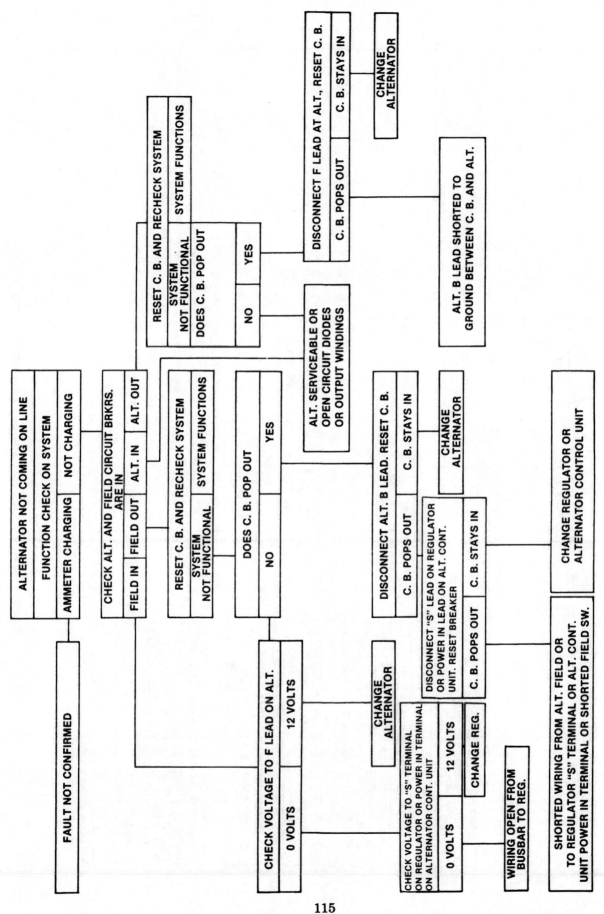

Fig. 2-50 Sample troubleshooting flow chart for an alternator system

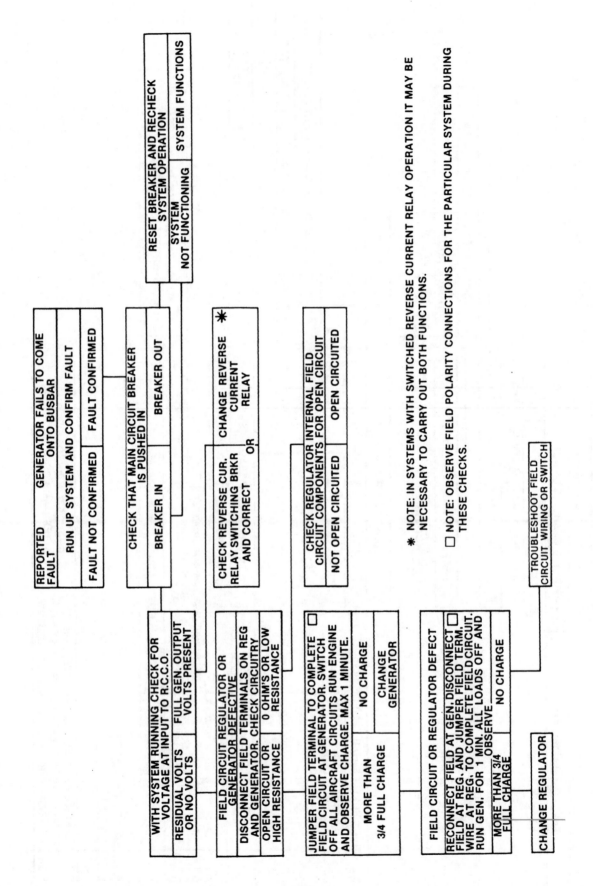

Fig. 2-51   Sample troubleshooting flow chart of a DC generator system

116

(c) failure of the reverse current relay or its associated circuitry to switch the generator output to the busbar.

Fig. 2-51 is a sample of a general troubleshooting procedure for a generator failing to come onto the busbar.

If we have a similar one, or both generators are not coming onto the busbar, or one generator taking too much of the applied load, some of the significant events which could cause these problems are as follows:

(a) a difference in the voltage settings of the two regulators.
(b) a bad ground on one generator causing an incorrect equalizing voltage at that particular equalizing resistor.
(c) a faulty voltage regulator.
(d) a faulty connection from the positive terminal of a generator.
(e) the equalizing coil circuit on the regulator system open circuit and a slight difference in regulator setting.

Most balancing problems will, in general, be solved by using the following procedures:

(a) run up one generator with approximately half load and allow the system to warm up for 10-15 minutes and stabilize.
(b) set up the system voltage to specification under the above conditions.
(c) close down the system and repeat operations (a) and (b) above on the other generator systems under exactly the same conditions.
(d) rectify any faults that occur during the above procedures which prevent achieving a proper result.

Review Questions:

1. What should be your first action when starting troubleshooting on a reported fault?

2. If you had a reported fault of all nav lights out and you traced the fault to all bulbs blown, what further checks are required?

3. You have a defect in the battery control system which is that the power remains on the busbar at all times. What would be two causes for this fault?

4. What are the two most common problems associated with twin generator systems?

Glossary

This glossary of terms is provided to serve as a ready reference for the word(s) with which you may not be familiar. These definitions may differ from those of standard dictionaries, but are in keeping with shop usage.

actuator A geared reversible electric motor used to give precise movements to controls and landing gear or landing lights.

ADF Automatic Direction Finder. A radio navigational aid that indicates an aircraft's bearing from an A.M. station.

ammeter The current monitoring device which indicates charge or discharge of the battery and is connected between the battery relay output and the electrical busbar.

alternating current, AC Electrical current which changes direction periodically and changes value constantly.

alternator A generating device which produces alternating current. In light single-engine aircraft, the alternator's alternating output is rectified into direct current by a diode assembly built into the alternator case.

alternator field switch The switch which controls the field excitation current in an alternator circuit.

auxilliary fuel pump See booster pump.

auxilliary power unit, APU A source of electrical power which is used to augment the aircraft electrical power systems for starting or ground servicing tasks on the aircraft. ·

avionics A composite word used in the aviation industry to define the area of *avi*ation elec*tronics*. Usually means electrical and electronics systems in the aircraft.

AWG Abbreviation for American Wire Gauge. Used with a number which identifies wire sizes in aircraft electrical systems; i.e., 20 AWG.

battery Electrical storage device that converts chemical energy to electrical energy.

battery master switch The switch which enables the pilot to select battery power "on" or "off." Its function is to energize the battery relay by completing the relay coil circuit to ground.

battery relay An electromagnetic relay switch which is used to connect the battery to the main electrical busbar. Also known as battery contactor or master relay.

battery relay closing circuit A circuit included in some aircraft which supplies power from an APU to enable the battery relay to close if the battery is flat and allow in situ charging of the aircraft battery from the APU.

brake solenoid A solenoid which is used to stop an actuator motor shaft and prevent overtravel when its limits have been reached.

booster pump An electrically driven fuel pump which is used to supplement the pressure delivered by the engine driven pump.

busbar A common electrical connection point. Busbars are usually identified by the type of equipment they supply, i.e., electrical or avionics or by their configuration of single or split.

carbon pile regulator A regulator which uses a stack of carbon washers as a variable resistor in a generator field circuit. Varying pressure on the pile varies field resistance.

circuit breaker A protective device which interrupts power to a circuit when the circuit's current rating is exceeded.

current The flow of electrons in an electrical circuit rated in units of amperes.

current limiter Part of a three-unit DC regulator box. Limits DC generator output to maximum rated output current.

current sensing coil Used in the current limiting unit to sense total current output from the generator. A current sensing coil is connected in series with a load or source and has a low resistance.

DC Abbreviation for Direct Current. Current which flows in one direction only.

de-energize To remove power from a relay coil and cause the relay to return to its normal state.

dimmer control Use on interior lighting circuits to raise or lower the level of illumination on panel and instrument lighting.

distance measuring equipment, DME An ultra-high frequency radio aid which measures distance to a DME vortac station.

down limit switch A microswitch which stops or indicates a mechanism when its down limit has been reached.

energize To apply power to a relay coil causing its contacts to operate.

excitation Result of applying current through a generator or alternator field coil and causing a strong magnetic field.

external power See auxilliary power unit.

field part of a generating device which provides the magnetic field for generation of electricity.

field switch Switch which controls the power of a field.

field windings Coils of wire in a field system which, when power is applied, creates the electromagnetic field of the generator.

fuse Circuit protection device consisting of a fusible wire calibrated to maximum current rating which, if exceeded, will rupture the wire and cut off power to the circuit.

generator A machine which converts mechanical power to electrical power. There are two basic types: a DC generator and a AC generator or alternator.

ground the negative side of an aircraft electrical system which in metal skinned aircraft is connected to the airframe structure.

ground power unit See auxilliary power unit.

hertz, Hz Frequency of an AC supply in cycles per second; i.e., 400 Hz.

inverter A machine which converts DC current to AC current.

loads Current consuming components connected to the aircraft distribution system.

microswitch A switch used in precise control applications such as landing gear limit systems. Named micro because a small amount of movement on the switch plunger will change the switch contacts over.

nav/com The navigation/communications transceiver used for air to ground and navigation on VHF frequencies.

navigation lights See position lights.

overvoltage unit or relay Detects overvoltage condition in an alternator system and open field circuit shutting alternator down.

position lights Lights to mark the extremities of the aircraft. Positioned on the tail and wing tips. Clear for the tail, red for the left wing tip and green for the right wing tip.

rectifier A series of diodes connected so as to convert alternating current to direct current.

relay Electromagnetically operated switch.

reverse current cutout An automatically operated relay used to connect and disconnect a DC generator to and from the battery.

rheostat A variable resistor used to control light intensity in instrument lighting.

rocker switch A switch which is operated by a rocking lever motion rather than a straight lever operation.

rotary inverter An inverter in which a DC motor drives a AC generator to produce AC current.

rotating beacon A red flashing hazard light on the fin of an aircraft to warn approaching aircraft of its position.

rotor The field component of a light aircraft alternator.

secondary power Refers to power which is not produced in the primary systems in the aircraft. The power produced by an inverter.

self-excited A generator which is excited by residual magnetism in its field system.

series coil A coil in the reverse current cutout which senses current flow in either direction.

shunt wound A method of connecting the field coils in a generator. The coils are shunted in parallel across the armature.

sinusoidal oscillator An oscillator that produces sine wave AC.

starter motor An electric motor which, when selected, connects to the engine gear box turning the engine over for starting.

starter relay An electromagnetic relay switch which, when energized, connects battery or APU power to the starter motor.

starter switch A switch which is spring loaded to "off" which, when selected on, energizes the starter relay coil causing the starter relay contacts to close and supply the starter motor with power.

static inverter An inverter that has no rotary moving components.

stator The stationary output windings in an alternator.

strobe light A light which is used as a warning beacon and is of extremely high intensity due to its use of high operating voltage and gas discharge tubes.

transistor A solid state electronic device which can be used to control the larger amount of current in its main circuit by applying small voltages to its control circuit.

transformer A device which can raise or lower AC voltage.

transponder A navigational radio aid which indicates to air traffic control, the position of an aircraft.

up limit switch A microswitch used to stop an acutator or mechanism which reaches its upper limit or to indicate when that limit has been reached.

volt ammeter A multi-purpose instrument, in fact, a millivolt meter, which by appropriate shunts, multipliers, and a selector switch will read both amps and volts on the one instrument.

VOM Volt OHM milliammeter. A multi-range meter which can measure various ranges of volts, ohms, and milliamps.

volt The unit of electrical pressure.

voltage sensing coil A coil in a relay which senses voltage applied. A voltage sensing coil is connected from a source to ground and has a high resistance.

voltage regulator A unit which controls the output voltage of a generator or alternator.

Answers to Review Questions

PART I SINGLE ENGINE

INTRODUCTION

1. The battery is used to start the engine and to operate electrical equipment if the generator should fail.

2. An engine driven generator provides power to supply all loads and charge the battery conserving the battery capacity for emergency use.

3. Auxiliary power is used to start the engine or ground service the electrical system in order to conserve the battery in cold climates or in short flight duration operations.

4. Electrical busbar provide convenient connection points for terminations, circuit breakers, fuses, and components in a distribution circuit.

5. The automatic resetting type.

6. Fuses are once only items and must be replaced if blown; hence the necessity to carry accessible spares in flight.

7. The master relay coil is grounded externally and has a relatively high resistance, while the starter relay coil is grounded to its case and has a relatively low resistance.

8. An actuator is used to give precise operating ranges to various mechanisms. This movement may be linear or rotary.

CHAPTER 1

1. The battery relay connects battery power to the busbar and to the starter relay.

2. The ammeter is connected in series between the battery relay output terminal and the electrical busbar.

3. De-energize means to remove the power from a relay coil and allow a spring to return the contacts to their normal position.

4. The special function of a starter switch is it is spring loaded to the "off" position.

5. Energizing the starter relay means that power is applied to the coil which creates electromagnetic force to close the contacts.

6. The starter circuit breaker also protects the instrument panel lights circuit.

7. The generator provides electrical power when the engine is running to charge the battery and supply the electrical loads.

8. The field switch is connected in series with the generator field coils and the field terminal on the voltage regulator.

9. The generator is excited by residual magnetism in the field system.

10. The voltage coil in the reverse current cutout.

11. The current limiter and the generator main 50 amp circuit breaker.

12. Output voltage is applied across the voltage regulator sensing coil. If the output voltage exceeds a preset value, the electromagnet effect of the coil opens the voltage regulator contacts. This inserts a resistor in the field circuit and reduces voltage output. The regulator contacts close, causing an increase in field current which raises output to the preset level. This operation is repeated rapidly.

13. By the reverse current action of the series coil in the reverse current cutout.

14. If current exceeds rated output value, the current limiting coil opens the current limiting contacts and inserts a resistance in the

field circuit. This reduces excitation and reduces output voltage. Reduced voltage means reduced current output. The depressed voltage is present as long as rated output current is exceeded. Therefore the current is limited to a safe value.

CHAPTER 2

1. The electrical busbar is the main interface between the power sources and the electrical loads.

2. In accordance with the normal rated current for each particular component.

3. A circuit breaker is a protective device to limit current flow in a circuit.

4. Split busbars are two or more busbars which are arranged so as to segregate delicate electronic equipment from power variations.

5. The split busbar relay has normally closed contacts. When starting or using ground power the relay is energized. This opens the split bus relay contacts and isolates the avionics busbar from the electrical busbar removing all power from the avionics busbar.

6. The circuit control switches are located on the front panel of the radio.

CHAPTER 3

1. Secondary power supplies are power supplies which have been converted from the main aircraft power source.

2. An inverter converts direct current to alternating current.

3. 400 Hz AC allows the design of lighter components.

4. The frequency is controlled in a rotary inverter by controlling the motor field excitation.

5. The voltage output of a rotary inverter is controlled by controlling the generator field excitation.

6. When the engine is running and the generator is on line.

7. A solid-state sinusoidal oscillator and a transformer.

CHAPTER 4

1. To save wire and to provide a good conductor.

2. No. 10 wire is large enough to handle the generator output current whereas No. 20 wire will handle only the field current.

3. The next higher current rated wire or the next lower gauge number, i.e., for 18 AWG use 16 AWG.

CHAPTER 5

1. No control over switching, no protection against reverse polarity connection.

2. By the use of a diode in the relay coil current.

3. The APU should be off, and the voltage should match system voltage.

4. The fuse will blow as the battery is shorted. The diode, D_2, prevents the battery power by passing the battery relay when the battery master switch is off and the APU is not connected.

CHAPTER 6

1. Its power to weight ratio is poor, low RPM performance is minimal, and servicing and maintenance costs are high.

2. Rotor contains field coils for excitation. Stator contains output coils for delivery of generated output. Rectifier converts generated alternating current to direct current.

3. An electromagnetic relay senses output voltage and opens its contacts. This inserts a resistance in the field circuit decreasing excitation, which in turn decreases output volts.

4. The field is excited by battery busbar voltage.

5. The lower contacts make with the moving contact and ground both ends of the field. This reduces voltage to zero.

6. The diodes in the output of the alternator block reverse current.

CHAPTER 7

1. The C170 battery is located on the port firewall of the engine compartment.

2. The battery capacity is 25 ampere hours.

3. Close to the battery box and relay.

4. This switch is designed to allow independent operation of the battery master switch, but ensures interdependent operation of the alternator field switch.

5. C150, C172 with Continental engines.

6. It is approximately 3″ square and its cover is secured by rivets. The cover is blue.

7. Heavy duty relays.

8. 1958 aircraft has a DC generator, a single busbar, a ground power system, and no relay control. 1971 aircraft has an alternator, a split busbar (automatic), spike clipping diodes, and a polarity controlled ground power circuit.

9. DA1—electrical busbar and ammeter. KA2—starter relay and starter motor.

CHAPTER 8

1. A functional understanding of the system. The ability to understand the wiring diagram. A logical approach to troubleshooting. The ability to use electrical test equipment.

PART II TWIN ENGINE

CHAPTER 9

1. By completing the battery coil circuit to ground.

2. Spring tension opens contacts A_1 and A_2.

3. To develop a millivolt drop proportional to battery circuit current.

4. Polarity and voltage must be correct.

5. Reduced voltage would be applied to the main busbar when the master relay is open.

6. The battery would discharge through the alternator field circuits.

CHAPTER 10

1. The ammeter would show standing loads plus starter relay coil current.

2. To allow the starting engine to fire on a correctly timed spark from the other magneto.

3. To provide an externally boosted source of power to the magneto primary coil to increase its spark intensity during starting.

4. Neither starter relay nor the booster vibrator would operate.

CHAPTER 11

1. The generators would not operate correctly in parallel operation and would not share the load equally.

2. 12.6V is the voltage of a fully charged battery. To ensure that the generator charges the battery, 12.6 volts or more is required from the generator to close the reverse current relay contacts and connect the generator to the busbar.

3. To correct pile pressure under varying temperature conditions.

4. By a leaf spring armature assembly.

5. It decreases.

6. It increases and decreases pile resistance.

7. It is wound on the same former as the voltage control coil.

8. Low resistance few turns.

9. .5 volts.

10. Carbon piles wear at different rates. Mechanical differences in armature tension. Deterioration of contacts.

11. More than .5 volts difference.

CHAPTER 12

1. Automatic connection of generator to busbar and disconnection of generator from the busbar under the appropriate voltage conditions.

2. The generator switch and the pilot contacts.

3. The generator voltage should exceed battery voltage by .35 to .65 volts.

4. 10 to 20 amperes reverse current.

CHAPTER 13

1. The regulator selector switch enables the pilot to select either the main or auxilliary regulator to control both alternator output voltages.

2. When the main circuit breaker for the failed alternator is opened by the pilot, this action also opens a microswitch contacts in the field lead to the alternator.

3. The ammeter would read left alternator output current.

4. The common feed to both field circuits is open circuited between pins B and F of the overvolt relay. This removes excitation from both generator fields and closed down both alternators.

5. Select the other regulator position on the regulator selector switch, and if the previous problem was associated with the previous regulator or overvoltage relay, the system should return to normal.

6. An RFI filter is designed to prevent electrical noise generated by the alternator from reaching the radio units and so causing radio frequency interference.

7. If one regulator fails, only one alternator system fails.

8. 0 volts.

9. It measures busbar volts.

10. System voltage
    0 voltage

11. An adjustment potentiometer is provided on each regulator.

CHAPTER 14

1. By a rheostat in the light circuit.

2. To balance the load if extra lights are installed.

3. Panel space was used up by using heavy rheostat as more lights were added.

4. The base.

5. The lamp load current.

6. It will decrease its resistance.

7. They would go to full brilliance irrespective of dimmer control position.

CHAPTER 15

1. Navigation lights. Landing light. Taxi light. Rotating beacon.

2. By arranging the light lens mask to cover 110° for port and starboard and 140° for tail.

3. By the use of a solid-state electronic flash unit which is in circuit when the selector switch is in "flash" mode.

4. A slipping clutch prevents extension over a certain airspeed.

5. A moving contact on the retract mechanism.

6. A split field series actuator motor.

7. Limit switches and a brake solenoid.

8. To balance the loading and prevent ammeter flickering.

9. 400V DC.

CHAPTER 16

1. One light is connected to each N.O. contact in parallel with the supply.

2. A warning horn operated by the throttle microswitch.

3. By the downlock levers coming into locked down position.

4. All gear legs must be locked up.

5. A reversible DC motor.

6. Down limit switch open or 15 amp circuit breaker popped.

7. To prevent horn blowing immediately up selection is made.

CHAPTER 17

1. A single pole double throw center off switch.

2. By the down and up limit switches.

3. The flap motor would continue to run and gear damage would result.

4. The flaps would neither go up or down.

CHAPTER 18

1. To provide a back up fuel pressure should the engine driven pump fail and for priming and transfer operation.

2. When throttle is advance to cruise or max RPM 1/3 above 1/3 open.

3. To detect loss of system pressure and automatically switch auxilliary pump to high pressure mode.

CHAPTER 19

1. Confirm the fault exists.

2. Check the system voltage with generator running to verify the system voltage is within limits.

3. Stuck master relay contacts. Internally grounded relay coil S terminal.

4. Failure to come on line. Failure to load share correctly.